MW00483859

Moving to Commercial Construction

by Stephen S. Saucerman

Included on the CD inside the back cover:

- A 300-page database of cost estimates, plus,
- An estimating program that makes it easy to use these costs,
- An interactive video guide to the *National Estimator* program,
- A program that converts your estimates into invoices,
- A program that exports your estimates to *QuickBooks Pro*.
- Blank copies of forms in the book in *Windows*™ and Macintosh formats.

The Show Me video guide to *National Estimator* starts a few seconds after you insert the CD in your CD-ROM drive.

Craftsman Book Company
6058 Corte del Cedro / P.O. Box 6500 / Carlsbad, CA 92018

Acknowledgements

The business forms featured in this book, and on the CD-ROM inside the back cover, were either created by the author, or taken from *Construction Forms & Contracts*, with the permission of the authors, Craig Savage and Karen Mitchell. You can order a copy using the order form in the back of this book.

The author and the publisher also thank the following companies and organizations for providing materials used in this book:

Associated General Contractors of America

Gilbank Construction

Westra Construction

Wisconsin Department of Workforce Development

Library of Congress Cataloging-in-Publication Data

Saucerman, Stephen S.
　　Moving to commercial construction　/　by Stephen S. Saucerman
　　　　p.　cm.
　　Includes index.
　　ISBN 1-57218-103-6
　　1. Commercial buildings--Design and construction.　2. Construction industry　I. Title.
TH4311.S28　　　2001
690'.5'068--dc21
　　　　　　　　　　　　　　　　　　　　2001028256

Contents

Chapter 1

The Change

Owners and managers of residential contracting businesses make scores of crucial decisions every day that affect the future of their companies — but none are so potentially risky as whether or not to expand their residential operations into commercial ventures. Commercial contracting (and the commercial building process) often seems to be an intimidating departure from the residential contractor's customary routine. And rightly so.

Taking this step involves considerable change. Suddenly, new and unfamiliar responsibilities become part of the contractor's daily building activity. Words and phrases such as *liquidated damages*, *bid bonds*, *contingencies*, and *prevailing wages* find their way into the conversation. Acronyms like *OSHA*, *HAZCOM*, *CSI*, *ANSI* and *ASTM* pepper the rather bulky *commercial project specification manual* (also new to the former residential contractor). They make it read more like a foreign-language code book than a building guide.

All in all, it can be an overwhelming plunge into a new area of expertise that's not quickly or easily grasped. But you've made up your mind: It's time to expand, and commercial contracting is the direction you feel you must take. So where do you begin?

Don't Worry, You're Not Alone

The anxiety you're feeling is natural, and completely justified. Commercial construction *is* a different world, presenting fresh challenges, new and more stringent procedures, and (if all goes well)

greater rewards. But, like most good things in life, success in commercial contracting comes at a cost. Fortunately, that cost will decrease as your knowledge and experience in commercial construction increase. In short, if you're going to get into the game, you're going to have to learn the rules and you'll probably also have to pay some dues!

I can relate to what you're going through. Before I ventured into commercial contracting, I spent many years in residential construction. I enjoyed the challenge that residential work offered and found the results rewarding, but there came a time when the routine (mostly new housing) became boring. I could feel myself growing stale.

At the same time, new competition was springing up every day and my profit margin seemed to be dropping rapidly. I was ready for a change, and commercial construction seemed the ideal route to take. It offered the opportunity for building on a grand scale, taking on an elevated occupational status, and, I was sure, an equally elevated income. Never mind whether this vision was real or not — that was how I pictured reality at the time. So I decided to make the jump, and I haven't looked back since.

Making the Jump

But it wasn't an easy transition. There was a lot to learn in a very short time. I was suddenly at the bottom of the experience food chain, struggling to climb up. Clearly, I needed to do some quick research into this new profession, but what I discovered was the first of the many surprises I would encounter in commercial construction.

You see, when I went searching for information to guide me into the business of commercial construction, there was very little to be found. Sure, the library had scores of books on construction itself, but few of these offered assistance with the *business end* of any field of construction — and virtually none offered that information on *commercial* construction. In fact, what little information I did find was more often than not:

▌ Thirty years old and sorely outdated

▌ Handyman and/or home improvement guides ("Changing the washer in your faucet" or "Building a backyard barbecue")

▌ How-to books for carpentry, framing, masonry work and so on

▌ Long-haired, scholarly construction-management guides written by people who *clearly* hadn't had much, if any, experience with a hammer

There seemed to be a missing link in the learning-how-to-do-it equation — and that's the gap I hope to fill for you.

What This Book Is About

This is a complete commercial-construction business reference, geared specifically to the average residential building contractor looking for a change. It offers today's residential general contractor, subcontractor, material supplier and designer a practical guide to making the complex transition from residential to commercial construction.

In the following chapters you'll find the real-life, common-sense guidance that I wish I had found when I made my switch. I had to learn the hard way, and I've made the mistakes — so you won't have to. I've also developed some creative and positive techniques that helped me make this transition manageable, profitable, and occasionally even fun. You'll find those here, too.

What This Book Is Not About

This is *not* a book about construction management. We will *not* dissect structural details, nor will I tell you how to make the picture-perfect concrete pour. Neither will we create seemingly-endless PERT charts, and (I promise!) you won't see the phrase "activity-on-node" even once.

I realize that construction management topics are interesting for builders, but there are already many excellent references available on these subjects. (You'll find a list from Craftsman Book Company in the back of this book.) And besides, they simply aren't relevant to our goal for this book. Instead of teaching you *how to build*, we're going to teach you how to move your current residential operation into commercial markets with as little disruption of your business, and your profits, as possible. We'll focus on the transition itself; how your goals will be altered, how your life will be changed, and how your horizons will be expanded.

But enough talk . . . it's time to get started. So, kick off your work boots, and get ready to *make that move!*

But First, a Few Ground Rules

I've made one important assumption while writing this book. I've assumed that you are (or were) involved in the residential construction or remodeling field and you have a fundamental knowledge and understanding of construction in general. This was necessary to avoid making the discussion too basic.

Also, I've opted to use "he" and "his" over "he/she" and "his/hers" in places where it's relevant. This is simply for the ease of reading, and definitely not meant as a slight to my female associates who make up a large and critical portion of the construction industry. And who, with the advent of thriving organizations such as the National Association of Women in Construction, will only continue to grow as an influence.

Finally, I've had to decide on a representative size (expressed in dollar volume of sales) for the commercial contracting company to target in this book.

This was necessary because running and analyzing a $5-million commercial construction company is significantly different from one doing $200 million or more, particularly when it comes to staffing, types of work, and administration. So I've chosen to base most of our discussion and examples on a commercial contracting firm that does approximately $5 to $20 million in annual sales.

I picked this particular level for two reasons. First, it's the dollar volume that I've worked with over the years, so it's by far the most familiar to me. Secondly, and perhaps most important, this is the dollar volume that I believe most of you will eventually fall into. You probably won't start off that high, more like $2 million in the beginning, but after a few years your company should be approaching the $5 to $10 million range. That's a realistic goal. I know far more commercial construction firms in the $10 million per year sales bracket than in the $200 million per year bracket. And, quite frankly, if you're doing (or have the ability to do) $200 million per year, you should be writing this book — not reading it!

> *"As technology advances and populations grow, the need for quality commercial construction grows accordingly."*

The Construction Industry

As technology advances and populations grow, the need for quality commercial construction grows accordingly. New businesses will be created, new manufacturing facilities will be built, and existing firms will expand, renovate, or rebuild operations to meet the ever-developing world demand. The work is out there. All you have to do is go after it.

Residential Contracting

Before we make that leap, let's define our players. We'll start with where you are now. As the name implies, *residential contractors* are those professionals who involve themselves primarily in aspects of residential building and remodeling — most often working with single-family homes. This group includes a variety of professional trades that are very similar to those in commercial construction, including general contractors (GCs), subcontractors, building material suppliers, and more.

Although the residential GC is the overall leader of the project, he must work with subcontractors, whose trades break down into subgroups of skilled and semi-skilled-workers. These subcontracted trades include plumbers, electricians, painters, carpenters, carpet layers, plasterers, excavators, and so on. All of the trades and material suppliers come together to form the team of residential construction professionals who create and maintain the homes and apartments in which we live.

Taken as a whole, the number of persons employed in residential construction is staggering. According to a recent U.S. Census, there are over 100,000 residential general contractors, who employ over 500,000 workers. Including single-family, multifamily and speculative contractors, they account for approximately 64 percent of the total general contracting market (Figure 1-1 A). On top of this there are legions of people working in residential subcontracting fields (shown in Figure 1-1 B), material supply, design, regulatory agencies, and related administrative fields.

This makes residential construction one of the largest employers in the U.S., if not the world. With so many participants, and so much power behind them, it's no wonder that they enjoy an abundant supply of information, assistance, and networking resources. There are residential builder's organizations, home design and construction publications, and even home-building and repair shows on public television. So why, if residential contracting has so much going for it, would anyone want to leave the fold?

The Bad News for Residential Contractors

Well, as it happens, this same abundance of participation in the field also creates an overabundance of competition and market pressure. There were times when I felt everybody I knew was a home

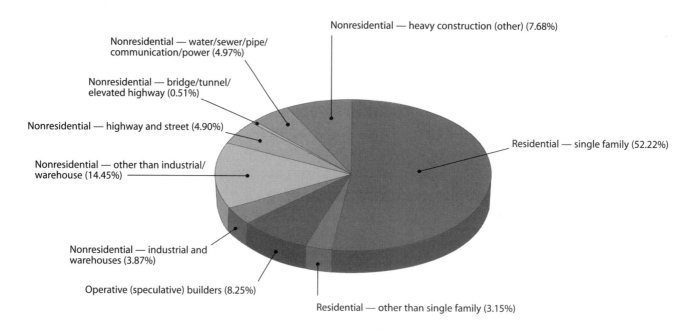

Nonresidential — heavy construction (other) (7.68%)

Nonresidential — water/sewer/pipe/
communication/power (4.97%)

Nonresidential — bridge/tunnel/
elevated highway (0.51%)

Nonresidential — highway and street (4.90%)

Nonresidential — other than industrial/
warehouse (14.45%)

Nonresidential — industrial and
warehouses (3.87%)

Operative (speculative) builders (8.25%)

Residential — single family (52.22%)

Residential — other than single family (3.15%)

Source: U.S. Census

A U.S. general contractors

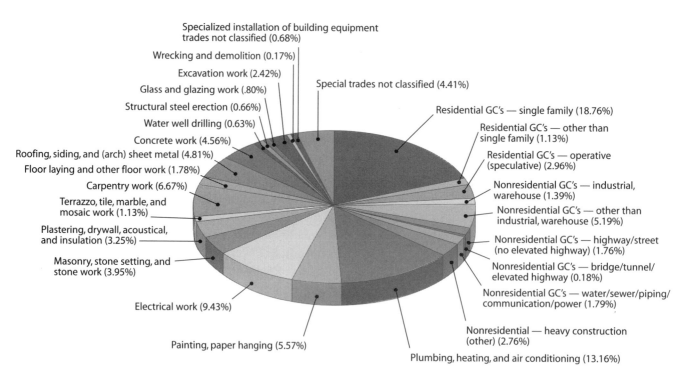

Specialized installation of building equipment
trades not classified (0.68%)

Wrecking and demolition (0.17%)

Excavation work (2.42%)

Glass and glazing work (.80%)

Structural steel erection (0.66%)

Water well drilling (0.63%)

Concrete work (4.56%)

Roofing, siding, and (arch) sheet metal (4.81%)

Floor laying and other floor work (1.78%)

Carpentry work (6.67%)

Terrazzo, tile, marble, and
mosaic work (1.13%)

Plastering, drywall, acoustical,
and insulation (3.25%)

Masonry, stone setting, and
stone work (3.95%)

Electrical work (9.43%)

Painting, paper hanging (5.57%)

Special trades not classified (4.41%)

Residential GC's — single family (18.76%)

Residential GC's — other than
single family (1.13%)

Residential GC's — operative
(speculative) (2.96%)

Nonresidential GC's — industrial,
warehouse (1.39%)

Nonresidential GC's — other than
industrial, warehouse (5.19%)

Nonresidential GC's — highway/street
(no elevated highway) (1.76%)

Nonresidential GC's — bridge/tunnel/
elevated highway (0.18%)

Nonresidential GC's — water/sewer/piping/
communication/power (1.79%)

Nonresidential — heavy construction
(other) (2.76%)

Plumbing, heating, and air conditioning (13.16%)

Source: U.S. Census

B U.S. construction work force

Figure 1-1
Construction breakdown by trade

builder or remodeler. Competition came out of nowhere. It often seemed that I belonged to a group where it was simply *too* easy to become a member.

In fact, in my market area, we had a far-too-frequently reoccurring phenomenon. Whenever the workers in our local auto plant (the primary employer in our town) went on strike or suffered layoffs, a disproportionately large number of new residential builders would suddenly appear on the scene. They had names like *B & D, S & S* or *B & J* — always follow by the term *Builders*. Those of us who were in the business for the long run assumed that the two initials belonged to two former auto workers who decided to join forces and become builders following the latest layoff.

Few were licensed (often, it wasn't required), and many were totally inexperienced. Yet, *there they were.* Of course, many of them merely flickered and then died away, but not before they managed to inflict a considerable amount of damage on those of us remaining in the local industry.

They would bid a half-dozen jobs at a cost well below their established competitors. Their clients — who thought they were getting a great bargain — greedily accepted the low bid with little or no investigation into the builders' background. You can probably guess the rest. Their projects inevitably floundered, the work was of poor quality or went uncompleted, another company had to be called in to finish the work, and *B & D Builders* was gone.

But the damage to the local market had already occurred. They had created expectations for an artificially low price among homeowners and homebuyers. A few of the established companies would try to match the prevailing rates and increase their volume to make up for their decreased profit margins — and that would further dip into our pool of prospects.

Desperate for sales, we'd find that we not only couldn't increase our profit margins, but we were actually having to *decrease* our markup just to keep up with the pack. It doesn't take long until your profits are no longer covering your overhead, and you're on the verge of following *B & D Builders* into

oblivion. Some of the companies survived this scenario, but many didn't. If you've been in this business long, you've probably seen this happen more than once.

Moving a Little Closer . . .

So you look to commercial construction. You see less competition in commercial work, and (most of the time) greater dollar volumes. Assuming that the profit percentages between residential and commercial are similar, a little quick math leads you to the rationale that *"as long as you're going to expend the energy, why not take your profit percentage out of, say, $2,000,000 rather than $150,000?"* And with fewer competitors, you think that you just might be able to find a *niche* (a singular market segment) that will satisfy your financial and business needs. With this in your mind, you find yourself one step closer to making your move.

Commercial Contracting

Commercial contracting is generally described as the *construction of any project of a nonresidential nature* — which is a pretty broad definition. It basically excludes single-family homes and rental properties up to about four units (depending on who you're talking to). It includes an almost limitless number of categories such as institutional, medical, manufacturing, civil, and so on. Figure 1-2 shows some of the common commercial construction categories.

Commercial General Contracting vs. Construction Management

One of the first decisions you'll need to make about commercial contracting will be your overall business philosophy. You need to decide on what approach to take in your new market. Do you want to be a:

▮ *commercial general contractor (CGC)*, or

▮ *construction management company (CM)*?

Although there are specialized situations in commercial work where a subcontractor is the lead or *prime* contractor (which we'll discuss later), a CGC

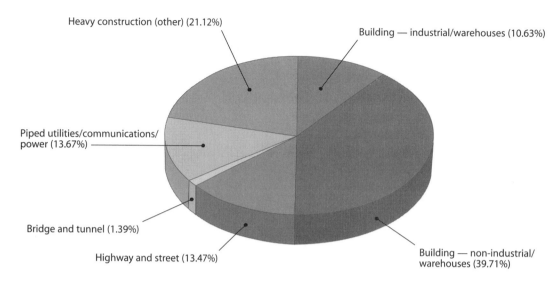

Heavy construction (other) (21.12%)

Building — industrial/warehouses (10.63%)

Piped utilities/communications/
power (13.67%)

Bridge and tunnel (1.39%)

Highway and street (13.47%)

Building — non-industrial/
warehouses (39.71%)

Source: U.S. Census

Figure 1-2
Work specialty for commercial contractors

or CM is generally hired to head up commercial construction projects. The differences between these two can seem a bit murky at times, so let's take a closer look at them.

Commercial General Contracting — A commercial general contracting firm is the more familiar type. They're often family-owned and run, sometimes two or three generations deep. They may employ three to seven full-time staff or family members in the office, as well as a few full-time tradesmen. The tradesmen are often carpenters who double as job superintendents, but they could also be cement masons, bricklayers, or excavators, depending on the CGC's specialty. The company may also keep additional trade employees on staff to make up crews for work that the firm does "in-house" (work that they don't subcontract out).

This group of full-time employees (those enjoying employee benefits, such as insurance, retirement, vacation, etc.) forms the nucleus of the organization. The rest of that contractor's work is completed with the help of subcontractors, suppliers, temporary workers, and other outside *independent contracting firms* (those providing their own benefits).

The number and type of key personnel is one notable difference between the CGC and the CM. Another difference is how they obtain their work. Although many older and more established CGC's enjoy a good percentage of *negotiated work*, most of their work comes from the *competitive bid process*. This is especially true of newer firms that haven't achieved the name recognition or clientele of an older, family-owned business.

We examine both the *negotiated* and *competitive bid* processes in depth later on in the book. For now, here's a brief look at the competitive-bid process.

▌ An owner hires an outside architect to draft the plans and specifications for the project and assist with preliminary budgets and the structuring of the bid process.

▌ A number of CGCs (usually three to six, but sometimes more) are invited to submit blind, competitive bid proposals for the project, based strictly on the plans and specs.

▌ The CGC solicits and receives subcontractor and supplier bids. These may or may not be marked up by the CGC, and then incorporated into the CGC's proposal, which is delivered to the owner by a predetermined date and time.

▌ The lowest qualified bidder is normally selected for the project. However, the owner almost always retains the right to choose the next higher bidder(s) if the owner feels he would receive better value for his money from them. (This could be for reasons ranging from a previous bad experience with a particular firm to simply feeling more comfortable with a more experienced contractor.)

▌ Once the project starts, the CGC and the owner usually communicate through the architect. The architect, although he primarily has his own interests at heart, acts as the owner's agent — and the owner must trust his judgment. The owner may or may not have established a trusting work relationship with the contractor, and often he has very little to say about what subcontractors and suppliers are chosen for his project.

▌ The CGC oversees the project and coordinates all the subs, suppliers, and equipment. Most likely, one of his own employees is the superintendent or job foreman. Depending on his specialty and the nature of the project, the CGC may perform some of the work with his own crews.

▌ If all goes well with the project and there are few problems, an end-of-the-job punchlist is worked up by the architect. The CGC completes the list, the final payouts are made, and everyone is happy and goes their separate ways.

Construction Management — In a construction management situation, the construction manager (CM) approaches the business a bit differently:

▌ The construction management company may be selected by the owner, or a few companies may bid on the job. If it's competitively bid, each CM will submit a proposal for the job which includes company information and a fee percentage. The fee percentage may or may not be the determining factor in who is eventually awarded the job, but it's a large factor.

▌ The CM who's awarded the contract acts as more of a fiduciary/agent to the owner and remains responsible for most all phases of the building process, including bid solicitation, job management, and accounting. Of course, this requires the CM to gain the owner's trust. That usually happens only after years of experience working with owners and/or very good public relations skills.

▌ Architectural services may or may not be supplied by the CM. Often, the owner will bring his own architect to the table. The CM works with both the owner and the architect. He will work out the overall scope, vision and schedule of the project with both the owner and the architect, but depend on the architect to supply the daily details. The CM and the architect work together to keep costs in line.

▌ Generally, the CM doesn't employ any full-time, permanent tradespeople. Virtually all of the construction work is done by independent contractors.

▌ The CM works directly with every subcontractor (who act as individual prime contractors) to maximize cost savings to the owner. Separate contracts are issued to all of the individual trades, and the CM oversees all the work. Sometimes the CM hires a general contractor to oversee the trades, and the CM oversees the GC. This hierarchy of responsibility is used on larger projects.

▌ In theory, the owner saves the markup taken by a CGC on the subcontractor and supplier quotes, but, of course, that savings must be weighed against the management fee paid to the CM.

▌ CMs are generally hired as watch dogs for large-dollar-volume construction projects where the economics of the arrangement make good fiscal sense. This is especially the case when there's a general contractor overseeing the job as well. In a large-dollar-volume project, there's generally room for both a general contractor and a CM fee.

Drawing Conclusions

In short, the difference between the Construction General Contractor and Construction Manager is more about the managerial arrangement and the relationship between the owner and the contractor, and less about the actual techniques of building the construction project itself. The same mason or electrician will more than likely be working on the job regardless of whether the lead player is a Construction General Contractor or a Construction Manager.

This comparison should give you some help in making your decision on how you want to define your new role in the commercial market. Of course, you won't be able to just grab onto this idea and use it. It's more likely to be a gradual decision, molded over time by your own experiences.

Chapter 2

Staffing the Commercial Construction Office

For the purposes of our discussion, let's say that you've chosen to become a Commercial General Contractor rather than a Construction Manager. This is the more typical route to take into commercial construction. Now it's time to develop the *nucleus* of your commercial construction office — those people who'll make up your core decision-making and administrative team. Since you've been in business as a residential contractor, you could already have some of these people in place. However, let's assume that you're starting from ground zero.

Weighing Your Needs

First, you must decide on the possible size and scope of your new commercial enterprise. This includes how many people you'll need as your nuclear office/administrative staff. These decisions should be based on:

▌ Your proposed volume of work

▌ Your anticipated generated revenue

▌ Your perceived role in the company

▌ Your immediate need for particular skills

▌ The available labor pool

Your Proposed Volume of Work

Do you perceive your new commercial company as a small- to medium-size venture with sales leveling out at between $3 to $5 million per year, or do you think you have the opportunity to achieve sales of $15 to $20 million within your first year or two? As the dollar volume of business goes up, the need for a larger office staff increases proportionately.

Your Anticipated Generated Revenue

Having chosen a target sales volume, will the profit from the revenue generated cover your office overhead? Figure 2-1 shows an example of an overhead calculation sheet for a company doing $5 to $6 million per year. You can use this form, or one that you design to suit your own business, to calculate your anticipated office overhead.

Of course, since you have no history as a commercial contractor at this point, some of the input will have to be projected on the basis of your proposed income. However, by calculating your estimated overhead, you'll be able to make adjustments to your expenses as needed. One area where these adjustments can be made is in office staffing. You can decide, using your projected overhead, just how many people you can afford to have on staff, and what wage you'll be able to offer to get the skilled and experienced help you need.

Your Perceived Role in the Company

Do you want to be a *hands-on* boss, performing some of the duties such as estimating or project management yourself, or do you picture your role as more of an administrator? Perhaps you'd like to take on the responsibility of sales and networking, and leave the day-to-day aspects of the construction

ABC Contractors
1234 Hammertacker Way, Lienwaiver, WI 53555 • Office (608) 555-1234, Fax (608) 555-1235

Overhead Calculations

Date: 12/31/00
Gross income this year: $6,000,000
Gross Income last year: $5,000,000

	Last year	This year	Last year month average	This year month average
Owner's salary	55,000	58,000	4,583	4,833
Office personnel wages	106,000	108,200	8,833	9,017
Employer's payroll tax expense (on above)	9,010	9,197	751	766
Office rent	10,600	10,600	883	883
Utilities	3,180	3,266	265	272
Office supplies	4,322	4,682	360	390
Telephone, including business at home	7,800	8,450	650	704
Mobile phone/pagers	ABV	ABV	ABV	ABV
Advertising, all types	4,320	4,400	360	367
Legal, accounting and professional fees	5,500	9,500	458	792
Licenses and dues	1,200	1,200	100	100
Office equipment, lease and rental	1,800	1,960	150	163
Computer hardware and software	5,650	5,500	471	458
Insurance, liability and auto	8,500	8,750	708	729
Medical insurance	76,400	80,200	6,367	6,683
Travel, hotels & meals	3,200	3,800	267	317
Truck expense, gas, oil, maint./repairs	JOB COST	JOB	JOB	JOB
Associations and publications	1,200	1,500	100	125
Education and seminars	1,800	2,000	150	167
Business entertainment	2,100	1,350	175	113
Business meals	ABV	ABV	ABV	ABV
Bad debts, ___%	7,500	9,000	625	750
Small tools and equipment	JOB	JOB	JOB	JOB
Depreciation	8,800	8,350	733	696
Other: _____	7,250	7,600	604	633
Total:	331,132	347,505	27,593	28,958
Overhead %: Total overhead/gross income:	6.62%	5.79%	6.62%	5.79%

From *Construction Forms & Contracts,* Craftsman Book Company

Figure 2-1
Overhead calculation sheet

operation to the people under you. Of course, in the beginning, there's *virtually no way* that you can take yourself out of the day-to-day operations. But right now we're still discussing your vision for your company, which is important in establishing your potential staffing needs.

Your Immediate Need for Particular Skills

Evaluate your existing skills and the skills of anyone you've already selected to work with or for you. Are there glaring gaps in your expertise? Do you already have a loyal estimator who could, with some training, step up into the role of commercial estimator? Do you have a superintendent with commercial construction experience? Or are these areas you need to beef up with new staff? These are questions you need to answer when determining who'll make up your nuclear staff.

The Available Labor Pool

Many times an employee will come along who's just too good to pass up. Maybe you aren't looking for someone to take on this particular responsibility right now, but you snatch him up because you may not have another opportunity. Perhaps his exceptional talent will bring in more sales, or will help you to operate more efficiently. Can you adjust your staff responsibilities to make a place for this person? Call it *proactive hiring*. Sometimes, in staffing, the risks truly do equal the rewards.

I've found that there are no hiring rules firmly carved in stone. However, there are certain responsibilities in any commercial construction office that have to be filled by *someone*. These responsibilities can be performed by several different individuals, or by one person with many skills. You can start out with a few people who'll be responsible for more than one aspect of the business (such as an estimator who also does some project management, or a receptionist who can also do A/R and A/P). Just keep in mind that if you're going to succeed, whatever tasks aren't covered by someone else will fall on your shoulders.

A Typical Office

Let's take a look at a typical commercial construction office setup. That way you can get a better picture of the players on your team, and their associated responsibilities. Figures 2-2 and 2-3 show the office arrangement for two commercial businesses doing different volumes of work.

You can see that as a company grows and expands and the volume of work increases, the various positions become more clearly defined. The scope and weight of responsibility for each area becomes focused on the person who's best in that particular role. A commercial contractor doing upward of $200,000,000 per year may employ large numbers of people, each an expert in their own field. At that point, the company would most likely be divided into several departments, such as Accounting, Supervision, Engineering, Safety, Estimating, Surveying and Layout, Scheduling, Cost Engineering, Quality Control, Labor Relations, Contract Administration, Material Handling and Receiving, and so on.

Staffing Your Office

But let's come back down to earth and look at the players in your more typical ($5 to $15 million dollar sales volume) commercial construction office. For our example, you'll have a general manager — that would be you, the owner/CEO — plus a receptionist/bookkeeper, an estimator, a project manager, and a field superintendent. This is your team, and they make up your nuclear employee base.

The General Manager/CEO

The general manager (GM) provides direction for the company and oversees the administrative and field personnel under him. And as the owner, you have a genuine desire to see the company succeed. You're in charge of tracking the work from a business standpoint rather than from the day-to-day construction detail level.

Small Construction Firm ($1,000,000 volume)	
CEO/Owner/Manager	General management Sales and marketing Estimating Project management (all phases) Safety officer Oversee accounting functions and payroll
Estimator (optional)	Estimating and costing Contract administration Project management (all phases) Contract administration, purchase orders General office duties when required Job closeout

Secretary/receptionist	General office duties
	Contract administration
	Job closeout (assisting)
	Bookkeeper/ledgers (daily accounting —
	   often the major accounting such as profit/loss and balance sheets
	   are sent out to an outside accounting or bookkeeping firm)
	Office administration and office supplies

Figure 2-2
Office arrangements for a small firm

Medium Construction Firm ($1,000,000 to $10,000,000 volume)	
CEO/Owner/Manager	General management and long-range planning Sales and marketing Some estimating and project management Contract administration Oversee accounting functions and payroll
Estimator (as volume increases, another estimator may be added)	Estimating Design/build and negotiated proposals Job buyout and negotiation Shop drawing and submittal coordination Contract administration May do some project management
Project manager or Job superintendent (as volume increases, another PM may be added)	Oversee field operations — may or may not be the actual job superintendent, depending on situation and need Safety officer May do a small amount of estimating Job closeout (field) and punch list

Secretary/receptionist	General office duties
	Contract administration
	Bookkeeper/ledgers (daily accounting) — may also assist in major
	   accounting duties such as profit/loss and balance sheets, but an
	   outside accounting firm would likely still be involved
	Office administration

Figure 2-3
Office arrangements for a medium firm

As GM, you'll watch over all facets of the business, including sales and marketing, accounting, estimating, project management and scheduling, contract administration, employee relations, and whatever else might come up. You're the one who negotiates new work contracts with architects and owners. When the work is awarded, you must read and understand the contract, whether it's a standard AIA (American Institute of Architects) prepared agreement or something drawn up by the clients' lawyers. It's your job to address any legal details that reflect positively or negatively on the company.

You must also work with your accounting firm and lending institutions to develop long-range strategies and future fiscal projections for the company. These goals inevitably require putting together profit and loss statements, balance sheets and other financial reports. You may also work on developing a company operations manual, and writing and setting up employee training procedures. Mediating employee grievances and handling legal issues and lawsuits are your responsibility. You do the majority (if not all) of the hiring and firing for the company, and have the final decision on all of the upper-level company policy matters, such as insurance, company benefits and public relations.

You may have to delegate the day-to-day technical work details to your staff while you cultivate sales opportunities by networking with local business leaders and decision-makers. That is, of course, when you don't have to come to the aid of your employees! It's quite common in small- to medium-size companies for the GM to be called upon to assist with estimating, project management, or even accounting when the pressure on others becomes too great. That's just one of those executive *perks*! Plus you'll be the highest paid employee in the company (although maybe not right away).

The Receptionist/Bookkeeper

The receptionist/bookkeeping position in a commercial construction office is a fast-paced, challenging one, with a wide range of duties.

First and foremost, the receptionist must answer all incoming telephone calls. To many people this responsibility may seem trivial, but I assure you, it's not. It may possibly be the most important sales tool you have. The first contact that many people have with your company will be by phone, and it's that first impression that'll set the mood for the business to come. A receptionist with a pleasant phone demeanor, who can listen attentively to your customer's problems, answer questions intelligently, take good messages and remain helpful and cheerful all through the day — even when under fire — is a very *skilled* individual. Unfortunately, these attributes can't be quantified, so you never really know their true value.

I will tell you, however, that I have chosen *not* to deal with companies whose employees have been rude, apathetic, or indifferent to me on the phone. In a situation where every other aspect of the company's work was equal to or even better than another, I went with the competition rather than be treated rudely. This "equality among companies" is exactly why a good, positive impression on the phone is so important. When the services are relatively interchangeable, attitude and a can-do approach may be all that separates you from the competition.

I don't know about you, but I want my company to take full advantage of every opportunity to make a sale. And you won't get those opportunities if the first voice your potential customer hears creates a negative impression. So choose your receptionist carefully.

The other duties of the receptionist/bookkeeper include:

▌ Tracking the plans and specifications needed for bidding

▌ Assisting with submittal and shop-drawing collection and submission

▌ Daily A/P and A/R ledger entries (the detailed accounting tasks would be done by an outside firm)

▌ Handling all daily mailings and overnight packages (which are plentiful)

▌ Ordering office supplies

▌ General office administration work

It's a lot to keep up with, and for their labors, receptionist/bookkeepers often make the lowest wage in the office. *Think very seriously before you fall into this pattern.* Paying your receptionist (or anyone else) less than their true value to your company will eventually cost you much more than you save. Once someone gains enough experience to realize their worth, they'll find another employer who'll pay them more — and you'll lose a valuable asset to your company.

What's worse, that person's experience doesn't just disappear from the market. It's very likely that your competitor will pick up on this "instant expertise," and while you're trying to find and train a replacement, they'll have an already-trained, highly-skilled person running their front office. You don't have to be Lee Ioccoca to see that this is hardly an efficient and economical way to run your business!

The Estimator

The estimator, as the title implies, is responsible for performing the bulk of the office estimating duties. His talent lies in his ability to accurately calculate the total costs for a variety of building projects. Generally working from a prepared set of plans and specifications, the estimator uses the following information to organize a comprehensive, competitive bid proposal:

▎ Supplier and subcontractor proposals

▎ In-house material and labor estimates and takeoffs

▎ Historical data (such as labor hours from similar completed projects)

▎ Productivity reports from previous jobs

▎ Current industry trends that offer a basis for arriving at future costs

Possessing the ability to compare outside subcontractor and supplier quotations is a particularly important skill. They're critically important to the proposal, and yet these quotes often arrive in a less-than-complete form. The estimator needs to be able to figure out if he's comparing apples to apples or apples to oranges: *Does this HVAC quote have gas piping? Did this plumber include all the excavation and backfill for his sanitary line?* Questions like these must be answered before he can evaluate the *true* amount of the bid. A careless assumption here can cost the company a considerable amount of money. He has to have the bid prices clearly established before he can enter them into his general construction cost summary.

Besides preparing conventional (competitive bid or negotiated) estimates, an estimator's typical workweek might also include duties such as:

▎ Distributing bid documents and bid invitations to those subcontractors and suppliers whose pricing he needs to complete the overall bid proposal. This isn't as simple as it sounds. The logistics of getting plans and specifications out (and then back again) to as many beneficial bidders as possible can be quite involved and complicated. The receptionist/bookkeeper may help the estimator in this task by keeping a log of plans/specs coming and going. Figure 2-4 shows an example of a typical plan log.

▎ Reviewing and updating in-house labor and manhour figures. These are the costs that the company assigns for estimating and performing work with their own crews. For example, your company might assign a figure such as $0.28 per square foot as the *labor unit* cost to install plywood floor deck, or $1.50 per linear foot to install 2 x 4 wood blocking. The estimator keeps a record of these historical labor records to use in his bids. He may check these numbers against commercially-available estimating manuals.

▎ Attending prebid meetings. These meetings are held during the period when a particular project is out for bid. They allow the estimator to meet with the owners and designers at the site where the project is to take place. This provides the participants with an opportunity to walk the project environment itself so they can clear up any early and/or obvious questions, concerns, or details. Any changes or additions to the bid may then be put out on *addenda* (see Chapter 6) before the bid due date.

ABC Contractors

1234 Hammertacker Way, Lienwaiver, WI 53555 • Office (608) 555-1234, Fax (608) 555-1235

Plan Log

Date: _____

Owner: _____

Contractor: _____

Project name: _____ Number: _____

Print #	Company	Contact person	No. of sets	Deposit ($)	Date out	Date in	Phone number	Notes

File: PLANLOG

Figure 2-4
Plan log

From *Construction Forms & Contracts*, Craftsman Book Company

▌ Providing technical support before and after the fact to project managers, subs, suppliers, superintendents and even the owners. This may include clarifying or explaining details that were sketchy or conceptual during the original bid process.

▌ Preparing change orders (or *contract modifications*) for work already underway. This is often done in cooperation with the project manager and project representatives and is particularly important on larger jobs where there may be frequent changes. This added work must be costed out and presented to the owner for approval before it can be performed — sometimes putting the project schedule on the line. Quick estimating may be critical to the continuity of the project.

▌ Aiding in the preparation and presentation of what I call the *"dog and pony"* shows. These are construction presentations given for prospective clients. The goal of these presentations is to convince the client that your company isn't only the best, but the only one they should consider to work on their proposed commercial project. Larger companies often solicit proposals in this manner, and you may go up against one, two, or more competitors. These presentations can take a great deal of preparation and often include budget (or conceptual) estimates, renderings, prepared speeches, and multimedia tools such as charts, slide shows, maps, and/or demonstrations using media-presentation software such as Microsoft's *PowerPoint*®.

▌ Tracking awards and (along with the project manager) developing purchase orders and contracts for the subs and suppliers whose proposals have been accepted. This period of time, after the bid has been *let* (awarded), is often referred to as the "buyout" period. This is when the estimator locks in the subcontractor and supplier numbers.

▌ Other miscellaneous duties may include advising the GM on the types of projects to bid, organizing warranty or callback arrangements, helping the project manager with the final closeout procedures for a project (including the collection of written warranties, as-built drawings, operation and maintenance manuals, etc.), and other day-to-day office and administrative functions.

Like all of the team members in a commercial office this size, the estimator may need to help out in other areas, such as project management or even accounting. As the office grows, you may hire more estimators and break down the estimating duties into several levels of responsibility. You can use titles like *Senior Estimator*, *Chief Estimator*, and *Estimator Assistant* to spell out the chain of command.

The Project Manager

While the estimator tends to have most of responsibility through the award stage of the bid process, once the project is awarded, the burden shifts onto the shoulders of the project manager (PM). He takes over control of the construction project. Responsibilities between the estimator and project manager may often overlap, but for our purposes, we'll define the PM's duties to include:

▌ Tracking the shop drawing and submittal process. Working with the estimator, the PM arranges for and tracks the submittals and shop drawings. (We'll look at these in depth later on in the book.) This is a critical responsibility because many companies won't release orders for products and services until approved submittals or shop drawings have been reviewed, signed, and returned by the architect and general contractor. Since this can often take two to three weeks (especially the shop drawings), and the actual order and delivery of the item could take an additional four to twelve weeks, it's easy to understand how neglecting this process may affect the project schedule.

▌ Contract administration, which includes developing, with the aid of the estimator, purchase orders and contracts for those suppliers and subcontractors whose proposals have been accepted. This is sometimes referred to as *purchasing*. This is normally the point at which the PM takes control of the project, allowing the estimator to go on to bid proposals and estimates for other projects.

▌ Overseeing field operations and crews. This is arguably the PM's most vital role: ensuring that both his own crews and the subs' crews perform their work efficiently and per the plans and specifications. Quality control is a large part of his responsibility. This isn't always easy, given the wide range of personalities and temperaments he has to deal with. His job is often more a diplomatic exercise than a technical exercise. The PM may have to act as a crisis/dispute mediator to settle problems between the subcontractors working under him. As with the receptionist, exceptional "people skills" are a must. The value of a PM can often be measured by his ability to get the work done, and on schedule, without alienating any subs or lowering crew morale.

▌ Acting as safety officer. Larger companies often have a full-time safety officer whose primary responsibility is to make sure workers are in a safe work environment. But in smaller companies, this responsibility usually belongs to the project manager or the field superintendent. Safety on the job site is not only a good policy for your workers — it's good business policy. Safety violations carry fines, and *OSHA* is only too willing to charge the violators. *HAZCOM* programs should be in place and product Material Safety & Data Sheets (*MSDS*) should be on display in the office and at the job site, readily available for all workers to see. (We'll discuss these items later on in the book.) But fines aside, the welfare of the worker should always be a primary focus on the job. Injuries not only disrupt lives, they disrupt work, raise workers' compensation rates, and put you in jeopardy of lawsuits from the injured individuals. With all of these potential areas for loss, safety makes excellent financial sense.

▌ Creating and maintaining project schedules. Like costing, scheduling is an important element of a profitable construction process. The PM is in charge of tracking, maintaining, and updating the construction schedule as the project progresses. Many firms do scheduling on the computer, using *PrimaVera® Suretrack*, *Microsoft Project®*, *Harvard Project Manager®*, and others. These programs make it easy for the PM to periodically update, shift, and print current schedules at any given stage of the project. He can even track productivity and manhours, or break the schedule into phases or labor groupings.

▌ Conducting construction progress meetings — usually held at regular intervals (about every two weeks) during the construction process. They often involve all the members of the building team: the CGC, subcontractors, suppliers (as applicable), the architect/engineering firm, the owners (or their representative), consultants or inspectors as the need arises, and the PM. These meetings are held to discuss anything relevant to the success of the project, which often includes schedule updates, questions and/or clarifications of the plans and specifications, the mediation of problems, and, for the owner's representative, an overview of the project's progress.

▌ Performing closeout procedures and satisfying the project punch list. While the estimator is gathering up many of the documents required for closeout (this also could be the responsibility of the PM), the project manager is doing his part to bring the job to a successful completion by overseeing the punch list procedure.

A *punch list* is a list of items that must be performed in order for the project to be considered officially complete, and before the final payments are made. It's generally compiled by the architect, engineer and owner when the project is in its final stages — usually a week or two before the completion date. Items on this list often include painting nicks and scrapes, drywall damage, and other minor problems, but may also include larger items such as missing hardware, base molding, or something that may have been back-ordered or, for some other reason, left incomplete. The successful performance of the punch list is strategic because of its connection with the last payment — the one that usually contains the contractor's profit!

Field Superintendent/Job Foreman

The final member of your nuclear team is the one most unique to your particular firm. Your choice of this individual is often a reflection of one of the specialties of your business. Maybe you like to keep a carpentry crew on payroll. In that case, you'll want a field superintendent who's an experienced carpenter by trade. If concrete is your trade of choice, your superintendent might have come up through the ranks as a cement mason or finisher.

While the PM's responsibilities straddle the fence between field and office, the superintendent/job foreman's duties are entirely field-related. Depending on the size of the project and/or the firm, this person may or may not actually swing a hammer. That's determined by how much *true* supervision is required for the project. Generally, the smaller the project, the more likely it the superintendent will actively work on the job. On larger jobs, the super's duties may be entirely administrative. There may even be a clause in the bid specification that calls for a *full-time superintendent* (one who doesn't perform manual labor) throughout the course of the project.

Allowing for Growth

You may also want to employ a few full-time tradesmen to make up your field crews. It's not unusual for a $5 million company to keep three to six carpenters on payroll at any given time. Depending on your situation, you may have fewer or more players on your nuclear team than in our example. The important thing is to make sure all of the company responsibilities are covered. If you've done that, then your plan is a good one. As your business grows, you'll know when you need to adjust and make changes.

Independent Contractor vs. Employee

Another decision that you'll face early on in your commercial contracting career is whether to classify workers as *independent contractors* rather than *employees*. There's a big difference, and how you hire and classify your workers can have a major impact on your business. The classification determines whether or not the company is required to pay social security, Medicare and unemployment taxes on the worker's behalf. Under a new Internal Revenue Service guideline issued in August, 1998, employers may classify certain employees as independent contractors if they can apply these three general criteria:

▮ *The company has a reasonable basis for the classification.*

▮ *The company classifies all similar workers the same.*

▮ *The company files Form 1099-MISC for the workers.*

What exactly does the IRS mean by a "reasonable basis"? Well, as with many things governmental, the answer depends on a number of *guidelines* that you can use to determine if a worker fits into the *independent contractor relationship*. Here are the guidelines:

▮ The *worker* can decide when, where, and how he or she will work. The employer doesn't necessarily dictate hours or demand that the work be done on company premises.

▮ The worker is paid hourly or by the job versus a monthly salary.

▮ The worker creates an invoice and bills the company, just like a vendor does.

▮ The worker can provide product or services to more than one company at a time.

▮ The worker furnishes his or her own office supplies, equipment, tools, and training.

▮ A contract (verbal or written) governs how work is performed and how services can be terminated if necessary.

▮ The worker may hire his own assistants and has sole discretion in delegating work to those helpers.

If these items don't describe the relationship you have with your workers, then you probably have employees, not independent contractors. Keep in mind that not *all* of these factors need to be present for an independent contractor relationship to exist. If most of these criteria are present, you can consider a worker an independent contractor.

The Office Itself

Physically, the commercial construction office is similar to most office setups. You'll need adequate room for your staff members and the amenities that keep office morale positive, like a coffee maker and water cooler. There are, however, a few items that deserve a little more consideration.

First is the location of your office. If you can afford it, you may want to locate your office near the center of business activity in your town. What about near the city hall or country courthouse? That's handy for obtaining permits and government documents. Or, what about locating your office near a plan room? That way, plans and specifications that are out for bid are close by, and you may save on plan deposits, not to mention time.

Then, of course, you'll need computers, printers, modems, and possibly even x-y axis tables for estimating and plotting (depending on your scope of services and preference for estimating). You'll also need a dependable phone system with one or more fax lines — this is essential. Among the supplies you'll need are basic startup forms for your office. The sample forms in Figures 2-5 through 2-9 are handy for any construction office.

Take your time setting up your office. Think of it as the foundation for your business. The more thought and planning you put into your office, the more stable the environment, and the easier the office work will flow.

Interoffice Memo

ABC Contractors
1234 Hammertacker Way
Lienwaiver, WI 53555
Office (608) 555-1234, Fax (608) 555-1235

Date: _____
To: _____
From: _____
Subject: _____
Project number: _____
Project name: _____

Please:

( ) Review and recommend   ( ) Shop drawings attached   ( ) Approve

( ) Follow up and report   ( ) Return   ( ) For your information

Note:
_____
_____
_____
_____
_____
_____
_____
_____
_____
_____

Reply:
_____
_____
_____
_____
_____
_____
_____

File: INTEROFF

From *Construction Forms & Contracts*, Craftsman Book Company

Figure 2-5
Interoffice memo

Memo

ABC Contractors

1234 Hammertacker Way
Lienwaiver, WI 53555
Office (608) 555-1234, Fax (608) 555-1235

Date: _____

To: _____

From: _____

Subject: _____

Project name: _____

Project number: _____

_____

_____

_____

_____

_____

_____

_____

_____

_____

_____

_____

_____

_____

_____

_____

_____

_____

_____

_____

_____

_____

_____

_____

☐ Please reply   ☐ No reply necessary   Signed: _____

File: MEMO

From *Construction Forms & Contracts,* Craftsman Book Company

Figure 2-6
Memo

Fax Cover Sheet

ABC Contractors
1234 Hammertacker Way
Lienwaiver, WI 53555
Office (608) 555-1234, Fax (608) 555-1235

Date: _____

To: _____

Attention: _____

Fax number used: _____

From: _____

Return fax number: _____

Project name: _____

Number of pages including the cover sheet: _____

If any of these fax copies are illegible, or you do not receive the same number of pages stated above, please contact us immediately at telephone number: _____

Remarks: _____

_____

_____

_____

_____

_____

_____

_____

_____

_____

_____

_____

_____

_____

_____

_____

_____

File: FAXCOVER

From *Construction Forms & Contracts*, Craftsman Book Company

Figure 2-7
Fax cover sheet

Transmittal Form

ABC Contractors
1234 Hammertacker Way
Lienwaiver, WI 53555
Office (608) 555-1234, Fax (608) 555-1235

Date: _____

Company: _____

Attention: _____

From: _____

Project name: _____

Regarding: _____

We are sending you:

	Copies	Date	Description
( ) As requested	_____	_____	_____
( ) Attached	_____	_____	_____
( ) Under separate cover	_____	_____	_____

For your:

( ) Records   Memo: _____

( ) Use and information   _____

( ) Approval   _____

( ) Review and comment   _____

( ) Use and distribution   _____

Via:

( ) Overnight mail   Remarks: _____

( ) Mail   _____

( ) Hand delivered   _____

( ) Fax   _____

Telefax number: _____   Copies to: _____

Number of pages (including this page): _____   _____

Signed: _____

File: TRANSFRM

From *Construction Forms & Contracts,* Craftsman Book Company

Figure 2-8
Transmittal form

Employee Data Sheet

ABC Contractors
1234 Hammertacker Way
Lienwaiver, WI 53555
Office (608) 555-1234, Fax (608) 555-1235

Date: _____

Position: _____

Start date: _____

Interviewed by: _____

Employee name: _____

Last   First

Address: _____

_____

Telephone: _____

Social security number: _____

Birthdate: _____

In case of emergency contact:

Primary contact: _____   Secondary contact: _____

Phone: _____   Phone: _____

Address: _____   Address: _____

_____   _____

Emergency medical information:

_____

_____

_____

_____

_____

_____

_____

_____

_____

File: EMPDATA

From *Construction Forms & Contracts*, Craftsman Book Company

Figure 2-9
Employee data sheet

Chapter 3

Finding Commercial Work

Now that you have your office staffed, the next thing you'll want to do is search out commercial work opportunities. Since you're the new kid on the block, most of your work opportunities will come from competitive bids. These blind bids, as we mentioned in the last chapter, usually go to the lowest bidder. That's OK; the negotiated and design/build opportunities will come later as your commercial contracting company (and your work reputation) grow and mature.

Covering Office Overhead

Since competitive bids will likely be the main source of your business opportunities for awhile, *you'll need to get good at it*! Here's one thing that's true of all areas of construction: If you're going to survive, you need to generate adequate revenue to cover your costs and overhead. And now that you're into larger and more costly commercial projects, you're going to need even more.

An office staffed with three to six employees, plus yourself, can easily cost you $200,000 to $250,000 per year in overhead. Much of this will go to salaries, but it also includes general expenses like rent, maintenance, office supplies, telephone, advertising, marketing, and so forth. That's a lot of cash! To cover this annual outlay, and hopefully generate a little profit on the side, you'll need to continually seek out new work opportunities. You know from your residential work experience that you only make money in this business when your employees are

actually working. So the work opportunities you find must be sound, lasting, and offer a reasonable chance of generating solid revenue for your company.

Residential Bid Opportunities

In residential construction, your sales and bid opportunities may have come primarily through word-of-mouth, personal referrals and calls generated by local advertising and the Yellow Pages. That was probably enough to keep you going. In commercial construction, however, the projects are larger, more sophisticated, and more costly — and your competition has grown in skill and tenacity. The stakes are higher. Although there may be fewer companies competing for the business, the level of the competition is probably more intense than what you're used to. So where do you begin?

You go back to the basics. Experience has already taught you a couple of valuable lessons:

1. The only way to win work is to bid work — and lots of it.

2. You can't win every bid, but the more you bid, the more you'll win.

Assuming that you can be reasonably competitive in your commercial market, it follows that by increasing the number of jobs you bid, two things will happen. First, you'll get more in tune with your market. Second, your estimating skills will improve. Then, as your knowledge and your familiarity with your competition grow, it'll only be a matter of time before the law of averages shifts in your favor and the jobs start coming to your way.

Where do you find the bid opportunities you need? Let me explain one of the most notable (and surprisingly pleasant) differences between residential and commercial construction: the way you solicit and acquire work.

In residential construction, there isn't an organized network of residential job information available to the average contractor. Bid opportunities are sporadic and often out of your control. What bid activity you do generate is usually more a product of luck, sweat, and hustle than of premeditated planning or an organized effort. Things are about to change for you . . .

Commercial Bid Solicitation

In the commercial world there are a variety of professional, organized, and effective construction job information services and networks available to any commercial contractor who's in search of bid opportunities. There's probably even one near you. By taking advantage of this assistance, you can greatly increase your bid volume, and in turn, your incoming revenue.

What are these services? Let's look at a few.

Construction Reporting Services

Construction reporting services are private companies that publish periodic reports detailing the construction activity in your area. The reports tell of projects coming out for bid, projects recently awarded, and even work that may be just in the planning stages. A variety of relevant information is included in the reports, including the plan holders, owners, project description, architect, prime contractors bidding the job (so subcontractors know who to submit bids to), bid due date, project budget, and more. You can submit a bid if you're a prime contractor, and have your name added to the list.

The reports are updated every time they're published (about once a week) to keep the information as current as possible. This updating is particularly important to subcontractors and suppliers (like fabricators). They can check the latest report to see if any new CGCs are on the list, and make sure that everyone bidding gets their quotes before bid time. It also gives the prime contractors who are bidding an opportunity to see who their competitors are.

Negotiated jobs are sometimes featured in these reports as well. A prime contractor who already has a project secured may want to solicit quotations from subcontractors and suppliers for work on the project. Like the competitive bids for CGCs, these sub and supplier bids are often given a set bid date and time, after which awards are made.

Of course, these services generally cost money, usually in the form of a subscription for a specified period of time. Your price depends on the geographical areas to which you subscribe. The larger the area, the greater the cost. You can run up some bucks, but it may well be worth it — depending on your bid volume, and what other access you have to existing information. The profit on just one job you get through one of these services will probably cover your subscription cost.

Two excellent reporting services that I've used in the past are F. W. Dodge Reports and Construction Market Data, Inc.

F. W. Dodge

Perhaps a recognizable name to many readers, F. W. Dodge is part of the McGraw-Hill Company Construction Information Group, which includes *Sweet's Architectural Record, Engineering News Record Magazine* and the *Construction News Publishing Network.* The information contained in the Dodge Reports is drawn from a database of more than half a million private and public, local and national, new and renovation/retrofit construction projects in seven different stages of development.

F. W. Dodge spends over $80 million per year to collect information on the construction industry, obtaining their data through marketing research studies, government sources, and its own network of over 1,200 reporters who collect information on construction activity nationwide. Dodge data is available in both electronic and printed formats, with different delivery options for the user. They collect blueprints, specs, and microfilm plans for most major projects. They also sponsor their own plan rooms in most major cities around the country, with satellite offices in smaller cities. There's generally one within a short traveling distance of major projects where prospective bidders may go to view the plans and specifications, gather data, and/or estimate their project of interest.

Here's how to contact them:

Phone: (800) FWDODGE or (212) 512-3442, or

check out their Web site at: *www.fwdodge.com*

You can also e-mail them at dodge_cust_svc @mcgraw-hill.com and they'll put you in touch with the regional office nearest you.

Construction Market Data

Another construction reporting service is Construction Market Data (CMD), founded in 1982, part of the Construction Market Data Group. The Construction Market Data Group is a network of construction information systems including R. S. Means, Architects' First Source and the AIA/CMD Profile. CMD has also recently joined forces with the Construction Specification Institute (CSI). They've agreed to a five-year alliance with the hope that they can provide members and customers better services and further the advance of technology and quality in the construction industry.

CMD supplies market data in the form of productive leads, project reports, contact lists, market penetration analysis and sales evaluation reports, pinpointing a variety of geographical locations. Their weekly construction and civil bulletins are color-coded for easy project identification and also track projects through all stages of the construction process: planning, bid, negotiated, post-bid, and subcontractor award. CMD offers other services as well, including daily cards, bid hotlines (for checking last-minute changes), plan rooms, and a service called CMD Online. CMD Online is a powerful *Windows*™-based system, that you can access

through the Internet. It's updated daily to provide project leads and company data that may be downloaded into several different in-house construction management software packages.

Contact them at:

Construction Market Data
30 Technology Parkway South
Norcross, GA 30092
Phone (800) 992-7572
www.cmdonl.com

Building Organizations and Associations

Building organizations and associations are also potential bid-generating tools for the commercial contractor. No matter what your trade, there's probably an organization or association that you can join. General Contractors have the ABC (*Association of Building Contractors*) and the AGC (*Associated General Contractors*). Remodelers have the NARI, subcontractors have the ASA (*American Subcontractor Association*), the skilled trades and unions have their own memberships, and suppliers and manufacturers have their own groups. Most of these organizations have offices at the national, regional, and/or local levels.

Depending on the organization, they probably offer some type of newsletter or publication designed to keep the membership informed on events affecting their livelihood. They often feature news like market activity, productivity figures, labor relations, minority business enterprise/EEO requirements, ADA updates, safety, training, legislation, legal concerns, insurance, tax issues, and more.

In addition, they very often include information on bid activity and work opportunities. Like the reporting services, they may provide pertinent facts on projects such as the architect, owner, bid date, and how to go about obtaining plans and specifications. Many organizations sponsor their own plan rooms, allowing members to access bid documents. Some even allow members to check out plans overnight or for the weekend.

Construction Magazines and Periodicals

There are many excellent construction-related magazines, periodicals and publications that include information on bid activity and upcoming jobs. Whether it be through listings in the back of the magazine, information in the features, or something more formalized such as bid advertisements, there's always something of interest. You may even find a project close to you. Figure 3-1 shows a typical Invitation For Bids from a construction magazine.

What about the contractor who likes to travel? Well, there's something for you, too. Since many of these publications cover a larger, more diverse geographical area than the typical trade organization newsletters, it's not uncommon to see work opportunities far from home — in countries all around the world. Opportunity is where you find it. During my career, I've run into many a contractor who has done quite well in markets outside the U.S.

Specialty Publications

If you're a contractor who has a narrow field of interest, specializing in the construction of hospitals, jails, churches, or schools, for example, there may be a publication that's geared toward your particular specialty! On a recent Internet expedition, I was surprised at the number of specialty publications (available both online and offline) that cater to just one type of specialty construction.

One such publication was *Correctional Building News* (CBN). They boast a readership of over 10,000, offering timely information and feature articles on the latest techniques in correctional facility construction for architects, correctional facility managers and construction professionals. You can contact them at:

Correctional Building News
11 Greenfield Avenue
San Rafael, California 94901
(415) 460-6185 or Fax (415) 460-6288
Email: info@correctionalnews.com

And there are many more to be found. For instance, have you ever heard of *Church Construction Magazine*? You can find them on the Internet at *www.churchconstruction.com*. If you have a special interest, just get online (or contact your local library) and search for key words that describe your specialty, such as *medical facility construction magazine*. You'll very likely come across something that'll interest you.

Finding Work on the Internet

There are all kinds of Web sites where you can find individual advertisements. Some of these are posted by local and state governments, some by private companies and organizations, and others by individuals. To find them, try the following:

1. Go to the Main Page of your favorite search engine (*AltaVista, Yahoo, Excite, Webcrawler, Lycos, etc.*).

2. Enter the type of project you're interested in, such as *hospital construction* (or make it more specific if you want to narrow your search options, for example *Wisconsin hospital construction*).

3. A list of projects will be displayed. These are primarily individual advertisements put on the Web by the hospitals themselves. You can click on and explore any that interest you.

This also works well with other combinations such as *school construction* or *institutional construction*. It just takes a little experimentation. Here are the URLs (Uniform Resource Locators) for some excellent sites that I came across while searching for construction information. They each include information on bid activity.

University and On-Campus Projects

Almost every university and college has its own Web page. Just type in the name of the campus you'd like to research. It may take a little patience to get it exactly right, but eventually you'll get to the page that you want. Here are a few you can try to get a feel for what's out there:

www.facilities.utoronto.ca/pmdc/pmdcmain.htm

www.facilities.utoronto.ca/PMDC/PROJECTS/ Consnews.htm

www.vpfss.swt.edu/CampusConst/projects.htm

INVITATION FOR BIDS

NORTH COUNTY NEIGHBORHOOD DEVELOPMENT CORPORATION

Site:
1234/36 Northwestern Avenue
Minneapolis, MN 55408

The North County Neighborhood Development Corporation, a nonprofit corporation, will accept sealed bids to furnish all labor, materials, tools, equipment and services for renovation of 1234/36 Northwestern Avenue. Bids will be accepted until 1:30 p.m. on Thursday, September 7, 2000 at the offices of the North County Neighborhood Development Corporation, 247 First St., Minneapolis, Minnesota 55408. No late bids will be accepted. The owner reserves the right to reject any and/or all bids and to waive any informalities in the bidding.

If you have any questions concerning the work write-up or on bidding, or if you wish to visit the site, please call John at 555-6172.

Contractors will be expected to supply a Performance and Payment Bond and meet U.S. Department of Labor Prevailing Wage Guidelines and submit all associated paperwork. Contractors may also have to meet certain construction goals from the Office of Emerging Small Businesses and labor utilization goals from the Minneapolis Department of Civil Rights.

Figure 3-1
Magazine bid advertisement

BuildNET and BuilderNet

You can also access other resources for building and construction on the Internet. Now, some of these sites don't actually *list* individual projects, but they do offer excellent information that'll keep you updated on building trends and activity. They may even provide you with a chance (via chat rooms, bulletin boards, and e-mail) to network with other builders and construction professionals — and that's another great way to create opportunity.

Here are two general sites for construction resources. Try BuildNET (at www.abuildnet.com) or BuilderNet is at (www.buildernet.com). Or try www.buildernet.com/library/library.html to go directly to their construction information. Just click on the *project advertisements* box and it'll take you to the current listings.

Government Projects

First, try the *General Services Administration (GSA)*. The fastest way I found to navigate this government site is to go to www.gsa.gov.search.htm and search for a keyword such as construction. Here are some other URLs to try:

www.gsa.gov.htm

w4.gsa.gov/projects/

w4.gsa.gov/projects/newconstruction.htm

w4.gsa.gov/projects/renovations.htm

w3.gsa.gov/index/index.nsf

Commerce Business Daily (CBDNet) (via the U.S. Department of Commerce and the GPO) is an excellent site for finding government projects of all kinds:

cbdnet.gpo.gov/index.html

Most state and local municipalities have Web site postings for construction jobs as well. Figure 3-2 is an example of a fairly typical posting. Do a search for your town, county or state and "construction bids." You'll be surprised at how much information you'll find.

The *Department of Transportation (DOT)* is a great site for those of you searching for highway and transportation work.

www.dot.gov/

You can link to the Department of Transportation in your state and then to local projects. It may take a couple of links to get to the location you want, but once you're there, you'll have it. For example, this is the address for the DOT in Colorado:

www.dot.state.co.us/public/about/index.htm

StructNet

This site has general construction advertisements:

www.structnet.com

Other Avenues

Here are some other avenues you can explore:

▮ Don't overlook the local newspaper. It's quite common to see bidding for construction projects, especially for city, county or state government projects, advertised in the paper. Sometimes advertising is even a requirement of the bidding procedure — a bid must be publicized for a certain period of time before the bid deadline. You'll often find these advertisements on or around the page with the finely-printed notes from the last city council meeting, or near the legal notices regarding estates, lawsuits, and real estate foreclosures. Also look for a heading called Proposals/Bids in the classified ads.

▮ Here's an idea is for subcontractors and suppliers who are looking for more leads. When I first entered the construction business, I sold lumber and building materials. One trick I used to find prospective clients was to go down to the local building inspection office each Monday morning and check the permit logs. They're public records, available to anyone who's interested. Periodically, I'd come across a solid lead for peddling my wares. It took some fast action on my part to follow up on these — after all, if the permit was taken out, the project had usually been started — but it worked out for me more than once. A side benefit of this routine was getting to know the people in the inspection and zoning offices. They often passed along tidbits of information on jobs in the early planning stages or soon to be started.

DGS Construction Bids

The Maryland Department of General Services posts bids daily on the DGS Bulletin Board located at 301 W. Preston Street, Baltimore, Maryland 21201. Last updated 3/20/2000

Bid Due Date	Time	Description	Location	Procurement Officer/Project Administrator	Job No.
23-Mar-00	10:00 a.m.	Paving parking lots & roads	Various state and county parks	Julie Smith	P-000-001-2345 Class: A
28-Mar-00	10:00 a.m.	Replace roof	Pikesville Military Reservation, Baltimore County	John Doe	P-000-002-2345 Class: A
29-Mar-00	10:00 a.m.	Repairs to Hunting Creek Bridge	Cunningham Falls State Park, Frederick County	John Doe	P-000-003-2345 Class: A
30-Mar-00	10:30 a.m.	Install pre-fab restrooms, construct camp pads & stone parking lot	Warrior Mountain W.M.A., Allegany County	Julie Smith	P-000-004-2345 Class: A
04-Apr-00	10:30 a.m.	Pave exhibit pathways and parking lot. Regrade for storm water measures RE-BID	Historic St. Mary's Commission, St. Mary's County	Mary Jones	P-000-005-2345 Class: B
04-Apr-00	11:30 a.m.	Recreation center outdoor facilities RE-BID	St. Mary's College of Maryland, St. Mary's County	Bob Smith	P-000-006-2346 Class:E
10-Apr-00	10:00 a.m.	Electric upgrades, fishing pier	Choptank River, Talbot/Worchester Counties	John Doe	P-000-007-2347 Class: A
11-Apr-00	10:30 a.m.	HVAC upgrade, Building "G" RE-BID	Maryland State Police HQ, Pikesville, Baltimore County	Mary Jones	P-000-008-2348 Class: B
14-Apr-00	10:00 a.m.	Asbestos abatement in pipe chases	Hospital building, Metropolitan Transition Center, Baltimore City	Bob Smith	P-000-009-2349 Class: A

Figure 3-2
Internet bid invitation

■ Most contractors hate this one, but it won't hurt you to get out there and practice the age-old art of the cold call. This is where you walk in unannounced — to a manufacturing or medical facility, perhaps — and just let them know you're around. This may be difficult for some of you, but don't ever underestimate the power of social interaction. I've seen it work so often, I'm convinced this is a viable method of finding work. Everyone, no matter how staunch and businesslike they may appear, likes to work with someone who's friendly. It's simply human nature. Of course, you'll need to be ready to provide these people with a resumé and information about your company. And you'll need to appear professional. The person you seek in these situations is normally the *Director of Engineering* or the *Facilities Manager*.

Conclusion

Of course, there's no better way to secure future work than to build up a clientele of satisfied, repeat customers — but that's something that takes time, hard work, dedication, responsibility and a reliable business sense to achieve.

For now, take advantage of the help and services that are out there for you. Stay focused on your goal and don't get discouraged if doors don't open for you immediately. You'll bid a lot of work that you won't get. That's simply the nature of the animal. But don't panic. I've had unbearably long stretches where I didn't win anything, sometimes even several months! But these were followed by a succession of five or six jobs in a row — often enough work to make my sales for the year. We're in an unpredictable business and you have to be able to roll with it. No matter how good the times are, you'll probably always be wondering and worrying about where your next job will come from! There's no solution to *that*.

Chapter 4

Working With Architects and Engineers

One of the greatest changes for the residential contractor who moves into commercial work is the introduction of the *architect* and *engineer* into the everyday building process. This dramatically alters the rules of the game.

The Architect

Architects, of course, are those design professionals who create entire sets of working drawings and specifications for building and construction projects — often with little more than a vague concept from an owner accompanied by some scratchings on the back of a napkin. The architect must not only create a coherent set of documents that convey the structure, use, details, and aesthetics of a particular building project, but he must do it in a manner that's understandable, economical, and convenient for the reader. The instructions must be clear enough not only for the engineer to understand, but also for the production manager or superintendent to relay to the average tradesperson.

To be able to claim the title *architectural draftsman* or *designer*, the person must comply with a strict set of instructional and licensing guidelines. Accreditation may vary from state to state, but usually involves the following:

▌ A college degree from one of many accredited schools of architecture. This may require five or more years of study.

▌ The successful completion of a licensing exam in the state in which the candidate wishes to practice architecture.

▌ Completing an internship under the supervision of other licensed architects. This can be done before or after taking the exam, but completion of the internship is a prerequisite of licensing.

Once the requirements are met in one state, a licensed architect may request a license in other states by way of reciprocal licensing agreements among participating states.

The Commercial Contractor/Architect Relationship

The commercial general contractor's dealings with an architect can be quite difficult at times. Depending on the relationship that's established between the owner and the contractor, the architect may be either friend or foe. When the job has come through the *competitive bid* process, with the architect and owner on one team and the commercial general contractor on another, the relations between the contractor and the architect can become quite strained — even adversarial. They may have opposing profit motives, with the architect protecting his and the owner's interests, while the CGC protects himself and his company. Under these circumstances, both parties need to make the effort to keep communications going and to be fair in their dealings with the other. This gets easier as you become more experienced in these work situations. There needs to be some give and take on both sides, and the understanding that there's often more than one way to accomplish any task and still get it done right. That's not always your way, just as it's not always *their* way.

On the other hand, if the project is contracted under a *negotiated* (or *partnered*) agreement, with the owner, the CGC and the architect all working together as a team, the relationship can be friendly and mutually beneficial. In this instance, fees for all the players involved are fixed early on in the process, so there isn't a limited pot of money in the middle for the participants to fight over. Everyone enters the agreement on a relatively even footing, and the motivating force tends toward cooperating on a shared goal rather than greed.

Choosing an Architect

Let's assume that you want to enter into one of the more agreeable types of relationships with an architect. Suppose you have a customer who wants to do a design/build project (where you as the CGC control the entire building process, including design), so you need to choose an architect for your team. What do you look for in an architect? What skills and personal traits are important to ensure a successful combination of technical know-how and professionalism?

Let's look at some considerations — from both an owner and a contractor standpoint — that you can use in your search for the right architect for you and your client.

Be Selective

First, get recommendations from your contractor friends and business contacts. Then make a few inquiries and ask those recommended firms to submit a statement of their qualifications, a company brochure, client references, and any other background information that seems appropriate. You're not insulting them by requesting this material. If they're a reputable company, they'll be delighted to share this information with you. If they hesitate, I wouldn't pursue them any further. Also, before you make an appointment, ask if they charge a fee for the first consultation; some do.

> *". . . . don't be intimidated by the idea of working with an architect simply because (for now) it's unfamiliar territory."*

Second, don't be afraid to shop around! Make comparisons just as you would if you were going to spend money on any other big-ticket item like a car, a home, or a remodeling contractor. You and your owner/client are putting a lot of money on the line, so don't think you're obligated to work with the first architect you interview. Also keep in mind that just as in any other business, architectural/engineering firms are in competition with one another. They're often very willing to negotiate a deal, including fees. So don't be intimidated by the idea of working with an architect simply because (for now) it's unfamiliar territory!

Finally, there's one more important consideration. Many owners rely on their architects to help them select a commercial general contractor for a project. Most established A/E firms have had experience working with the building companies in their area and they can help the owner/client put together a pool of prospective bidders. This opportunity for referrals is a good reason to get to know and to cultivate a good working relationship with several different architects and A/E firms.

Consider Their Attitude

You'll notice that I list *attitude* second. That isn't by chance. An architect's attitude is extremely important. In your interviews, pay close attention to their outlook toward you, the owner, and your project. Then answer the following questions:

▌ Does this architect, and his firm in general, display a *genuine desire* to understand you and your client's particular situation?

▐ Are they making an effort to communicate clearly with you? Or, are they talking over your head and using so much architectural/technical jargon that you haven't a clue what they're saying half the time?

▐ Are you comfortable that this firm will provide you with a creative and energetic design effort and will ensure that your requirements and your client's wishes and requests are met? If, in the first meeting, they're already telling you *"how swamped they are"* or using phrases like *"perhaps we can squeeze you in here,"* then be wary. Ask yourself, *"If I'm sitting here at their desk and discussing a work proposal, why am I not the most important customer they have? If this is how they treat me now, how much time and consideration will they devote to my project, my client and me when we're working together?"* Remember, they're the one looking for a job, not you.

▐ Are they easy to talk to or do they act distant and superior? No, I'm not being petty. This is a common complaint among builders in the commercial construction industry. Some architects seem to confuse themselves with God, and any comments or suggestions a mere contractor might have are treated with disdain. I'm not implying that you have to be best friends with your architect; this is a business arrangement, not a partnership. But you do want to work with someone you feel is approachable. After all, you and your client may be spending a lot of time with this person and his firm.

If you feel dissatisfied with the answers to any of these questions, be cautious. You might even want to move on to the next candidate. There are usually plenty of architects you can choose from in your area. And, as in any occupation, there are good ones and bad ones. If you take your time, you'll eventually find one who suits your needs.

Scope of Services

Architectural/engineering (A/E) firms do far more than draft up plans and specifications. In the qualification package you request from each firm, you'll find a listing of the services that they offer. Make sure you understand which of these services will become a part of your own agreement.

Common architectural responsibilities include:

▐ Consultations

▐ Preliminary design

▐ Engineering (civil, structural, mechanical, and electrical)

▐ Interpreting regulatory requirements, building codes, zoning laws and covenants that affect the project

▐ Budgeting

▐ Field management (supervising work to make sure the project is being built according to the plans and specifications)

▐ Contract administration

▐ Preparation of change orders

▐ Project closeout (warranties, as-built drawings, etc.)

▐ Other specialized design work as necessary

Economy

Once you've relayed your ideas to the architect and the preliminary sketches are being worked up, ask yourself this: *"Is the architect keeping track of economy?"* Put another way, is he being cost-effective with your (or your client's) money? Early in your discussions, you probably came up with a rough construction budget (which, in my experience, always ends up being too low), and a tentative project schedule. One of the architect's primary responsibilities is to *realistically* get you (and the client) as much product as possible for your construction budget. I call this *maximizing your construction dollar*. At this point you may still be in negotiations with the architect, especially if it's a small firm. It may not be too late to pay him off (if necessary) and choose someone else if you feel that he's not as budget-conscious as you'd like.

Value Engineering — Maximizing your construction dollar is known in the industry as *Value Engineering (VE)*. Your goal is to save money but still retain a level of quality that's acceptable and in compliance with the plans and specs for the project.

Unfortunately, a common complaint among commercial builders is that architects frequently specify products that they pull out of building materials catalogs, recent trade show literature, or some nonstandard or prototype source, and often don't have a genuine feel for the actual industry standard. When it comes bid time, you find that these items aren't readily available or attainable. They usually end up having longer lead times, added freight charges, and disproportionately larger prices than a similar standard item. If it's a very specialized, single-source item, the manufacturer may not even be willing to provide a cost until the order is placed. That means you have to come up with a suitable allowance to cover the cost in your estimate. The problem is that even if you've covered your cost correctly, the item may not be attainable soon enough to make it a viable product for your project — and you'll eventually have to come up with a suitable alternative or delay the project. Generally the choice is to go with a suitable alternative (in the name of value engineering).

These situations are very frustrating for commercial construction estimators. You know that there are often suitable substitutions for these items, available locally for much less and with a far quicker delivery time. Not only could you save the owner the added cost for the item, but you could also keep to the established construction schedule, decreasing interim money costs. Finally, using familiar items reduces labor hours for installation and probably increases the number of subcontractor companies that are qualified or willing to quote the job.

In a negotiated bid process, a solution to this type of problem is available to you. You can submit a request to the architect to use a voluntary alternate, and include your reasons why this particular item should be substituted. If your reasons are good, the architect will allow the change — *before* it becomes a problem. The new item will become spec in an addenda to the bid documents. This is just one reason that most contractors prefer negotiated contracts.

Project Budgeting — And while you're discussing budgets with the architect, be sure to clarify early on *exactly* what the architect's responsibility will be in case the project comes in considerably over budget after you've bid it out to subs. Be specific. Make sure you understand this completely because — and pay attention here — *architects quite often design projects that end up going well over budget.*

In a design/build situation, the commercial contractor has the responsibility of working up the budget numbers for the team. This gives you a bit more control over the situation. In a true competitive bid, however, the circumstances can be quite different if the client/owner has hired an architectural firm that doesn't have good estimating capabilities. I've bid *many* projects that have come in so disproportionately over the architect's original budget that all the bids were thrown out. That's even though all the bids were similar in cost, signifying the *true* value of the project. The project would then have to be redesigned, scaled down, and put out for bid again a month later. The crazy part is that the architect's redesign is often *billed as an additional cost to the owner*. The owner has to pay more hourly charges to get the job done right! (Now why didn't I think of that last time I messed up on a job? I paid out of my pocket to put it right!)

The owner has wasted a large chunk of time, and has to stumble through another design and bid process. And it may not be the last. I just finished bidding a church addition project for the *seventh* time! We did end up getting the job, but there are two other CGC's out there who put an enormous amount of time and effort into this process and didn't get anything! And, even though this project hasn't even started, the owner and architect are already tangled up in a bitter argument over the continuing fees.

In fairness to the A/E firms, it's realistic to expect about a 5 percent deviation from a budget estimate. It can even go as high as 7 or 8 percent, but any higher than that, and I'd begin shouting for a new negotiation — *loudly*.

Energy and Facility Planning

A good architect will possess the ability to *look down the road* for his client. This means he'll design a building that's cost effective, not only initially but also down the road. It can show up in many other forms: traffic planning, expandable end walls for future growth, life-cycle evaluations, or electrical service that takes into account the automated line you plan to install in about three years. But energy efficiency is one of the most important areas of cost efficiency.

> *"A good architect will possess the ability to look down the road for his client. This means he'll design a building that's cost effective, not only initially but also down the road."*

By predetermining a building's potential energy usage, the mechanical engineer (who may or may not be a direct member of the A/E's staff) can form a reasonable idea of what the owner will likely pay in utility costs. Once these costs are determined, the owner can make an informed decision about whether he wants to increase the initial energy-saving attributes in the building. He can weigh the cost against the future benefit of adding more insulation, opting for higher-efficiency equipment or appliances, or utilizing solar energy, etc.

The final decision on whether to decrease (or possibly increase) energy usage will probably come down to the pay-back period and the owner's long-term plans for the building. When will the owner see a return on that initial investment? Does he plan to keep the building for five years or ten years? Will he reap the benefit of energy planning or can he pass the cost on to a new owner? It makes a difference.

Cost of Service and Fees

There isn't a standard fee for architectural services. Fees vary from architect to architect, and from project to project. They're usually the result of negotiation. There are, however, a couple of methods that you may find helpful when attempting to establish fees for your project.

The Competitive Bid — The first method is to simply have the A/E firms *bid competitively* on the job, just like contractors do. This would require you to work up a package with your needs and requirements, deliver the package to maybe three firms, and set a bid date. You can request bids from as many firms as you want — but if you have too many involved, that's likely to dampen their enthusiasm for your project. That's usually counterproductive.

A good rule of thumb is that you want enough bidders to give you a reasonably competitive price on your project, but still give those bidding the impression that they have a good chance for success.

Selection Based on Qualifications — The other method is to simply select a prime candidate based on their qualifications, and mutually negotiate a fair fee arrangement. I heartily recommend this selection procedure for a couple of reasons. First, it simply forms the foundation for a better relationship. *Competitive bidding*, by its very nature, creates an *adversarial relationship*. That isn't how you want to deal with an architect. Second, any cost savings that may result from the competitive bid will probably be small, especially when you take into account the entire scope of the project. Is a fee that's $1/2$ percent less than someone else's worth it? Wouldn't it be better to have an architect who's genuinely and sincerely behind your project?

I've observed from experience that a good architect who enters into a relationship as an equal will easily make up that $1/2$ to 1 percent difference in fees with value-engineering and cost-saving suggestions.

You Get What You Pay For — Unfortunately, or fortunately, depending on which side of the fence you're on, architectural services are no different from any other purchased professional service. *You get what you pay for!* In most cases, the more money you spend, the more complete and detailed the information you'll receive.

When you think about it, that may actually be a good thing. Since the needs of owners vary tremendously, they may not want to pay for full-blown architectural services. If they're building a shopping mall, that's one thing. But what if they're only building an unfinished storage facility or a garage? These projects are relatively straightforward and require little technical expertise. This flexibility in fee structure is just what that owner needs.

The Engineer

An architect is often assisted by a variety of engineers who help coordinate information on *structural, civil, mechanical* and *electrical* details for the design package. The architect may be the engineer as well, if he has satisfied all the educational and licensing requirements.

If you compare the goals of an architect and an engineer for a project, you might say that the architect wants the project to be attractive, while the engineer wants it to remain standing. The engineer is the one who ensures that the construction is structurally, mechanically, and electrically sound and correct. He's not worried about materials that clash in color or texture, or whether the building has poetic lines. He simply wants the building to work right!

Some architectural firms have their own engineering departments and keep full-time registered, licensed engineers on staff. Smaller outfits subcontract engineering services through independent engineering companies.

Engineering Qualifications

Becoming a Professional Engineer (PE) typically requires a college degree in engineering, an eight-hour *Engineer in Training* exam, four years work in an engineering position that offers an increase in engineering responsibility, and then passing another

eight-hour exam on engineering *Principals and Practice.* While some states have requirements that vary somewhat from these, every state strictly regulates the licensing of professional engineers.

Engineering fields are broken down into many different disciplines. Here are a few branches of engineering that you may be familiar with:

▌ Chemical

▌ Civil

▌ Electrical

▌ Environmental

▌ Hydraulic

▌ Marine

▌ Mechanical

▌ Metallurgical

▌ Production

▌ Sanitary

▌ Structural

As a commercial contractor, you'll primarily be dealing with four engineering fields: *civil, structural, mechanical,* and *electrical*.

Civil Engineering

Civil engineering (CE) is broad in nature and often incorporates many engineering fields. It generally includes the *structural engineer*, the *water resources engineer*, the *geotechnical engineer*, the *transportation engineer*, and the *construction management engineer*. Some of their duties include planning and designing governmental and private physical works facilities for use by the general public, city or regional planning, highway construction, and pipeline layout and construction.

Structural Engineering

A structural engineer (SE) works with a wide range of clients including private industry, management consultants, financial specialists, architects, quantity surveyors, local and central government agencies, legal professionals, environmental specialists, and builders and contractors like you. Structural engineers plan and design bridges, buildings of all kinds, and specialized structures like television and radio towers, radar stations and power plants. They're particularly focused on the structural integrity of the project. Any project that requires concrete, steel, aluminum or wood to carry the load must be have a SE working with the architect in its planning and design.

Mechanical and Electrical Engineering

Mechanical and electrical engineers (MEs and EEs) are involved with the plumbing, HVAC, and electrical aspects of building project design and construction. They work up load calculations, determine appropriate equipment size, do design layouts (often from floor plans delivered by the architect), and assist with permits and other code compliance issues. Their stamp (seal of approval) is commonly required on the drawings as part of the compliance procedure, and once stamped, those engineers become liable for the mechanical and electrical designs they approve.

Professional Organizations

The architectural and engineering fields both have organizations to back them up. The American Institute of Architects (AIA) helps create and publish many of the architectural/building forms, documents, and agreements used in commercial construction.

Engineers have many professional organizations, including the *American Society for Civil Engineers*, *National Society of Professional Engineers*, *National Council of Examiners for Engineering* and *Surveying and Truss Bridge Laboratory*, that work for the benefit and improvement of the engineering fields.

The selection of an architect or engineer isn't unlike the selection of any other critical professional service. To make an informed choice, you have to know the right questions to ask, and you also have to not be afraid to ask them! As in all professions, there are good and bad firms. Don't get caught up in the flashy window-dressing promoted by most of these firms, but rather concentrate on the particular needs of your project. This requires a careful investigation of each firm and a willingness to ask tough questions. Remember, their services are going to be costly. You want to get exactly what you need — not too much, and certainly not too little.

Chapter 5

Construction General Requirements

For many commercial contractors, general requirement cost overruns are a killer — and often mean the difference between a profitable job and a loser. They're particularly dangerous because they can crop up unexpectedly during any phase of the construction process. They can be underestimated, just plain missed during the estimating process, or they can get out of control if they're not closely watched.

What are general requirements? They are all the peripheral job costs that *aren't commonly included in the individual subcontractor, supplier, and in-house material and labor estimates and proposals.* Still with me? Good. That definition is fairly broad, but then general requirements are fairly broad.

General requirements are those job costs that tend to be difficult to define and categorize during the bid process. They're items that belong to the job and yet often don't fall *clearly* into any one subcontractor or supplier's scope of work. Figure 5-1 shows a typical general requirements worksheet. (The code column is simply a numbering system for accounting purposes based loosely on the CSI division 1 subcategories.) You can see that many of the items on this list, like toilets, traffic barricades, temporary fences and temporary electrical power, are fairly common to any job. The question is, who pays for all these expenses? If no one else picks up the tab, the answer is you! And, if you don't watch these costs very carefully, they'll suck the profits right out of your pocket.

I've been there — and I've been burnt. It's surprisingly easy to be lured into a "Don't worry, we'll-work-it-out-when-we-get-out-there" kind of mentality during the estimating stage. Even seasoned construction estimators and project managers, who're very skilled at addressing the more tangible direct costs like excavating, masonry, or electrical work, often come up short when addressing these not-so-tangible items. But, in spite of the difficulty, this general requirements tiger *can* be tamed.

Why Should You Care About General Requirements?

Let's look at an example of why you should concern yourself with general requirements. Suppose you're a commercial contractor bidding on a project, and you've just received a masonry quotation for the bid you're putting together. It brings your total masonry proposals to four. You set out to analyze each one and select the *qualified lowest bid*; the quote that you'll eventually include in your prime proposal.

Right away you notice that there's an enormous difference between the highest and lowest price. You examine the proposals further, and to the best of your understanding, it appears the lowest quote includes the exact same scope of work as the highest. Now, it's possible that it's true. The low bidder may just be "hungry" and really going after the job, which isn't unheard of in construction. Sometimes a

ABC Contractors

1234 Hammertacker Way, Lienwaiver, WI 53555 • Office (608) 555-1234, Fax (608) 555-1235

Field Project General Requirements Worksheet

Code	Description	Qty	Unit	Labor $/unit	Labor total	Material $/unit	Material total	Equipment $/unit	Equipment $/total	Line item total
1.01	Superintendent/foreman									
1.02	Field engineer/timekeeper									
1.03	Night watchman									
1.04	Mobilization (in & out)									
1.05	Layout & surveying									
1.06	Hoisting & cranes									
1.07	Job office/trailer/supplies									
1.08	Toilets/drinking water									
1.09	Water for construction									
1.10	Telephone									
1.11	Temp. light/power									
1.12	Temp. heat/enclosures									
1.13	Permits & general fees									
1.14	Sewer/water connect fees									
1.15	Photographs/marketing									
1.16	Inspection & testing									
1.17	Samples & submittals									
1.18	Temp. roads & access									
1.19	Temp. fence, protection									
1.20	Temp. safety enclosures									
1.21	Winter weather/snow removal									
1.22	Clean/wax floors									
1.23	Cleaning (windows/misc.)									
1.24	Special trucking/freight									
1.25	Trucks & maintenance									
1.26	Small tools & equip. repair									
1.27	Safety meetings/HAZCOM									
1.28	Project signs/identify									
1.29	Dumpster/trash removal									
1.30	Site & street cleaning									
1.31	Protect finish work									
1.32	Traffic control/barricades									
1.33	(Other)									
1.34	(Other)									
	Totals									

File: FLDREQ

Figure 5-1

Field project general requirements worksheet

sub will take a job for little or no profit just to keep his crews busy. But then again, that inordinately low quotation *may not have included money for their general requirements*.

So you call the low bidder to discuss the project. Sure enough, you find that he hasn't included any of the winter weather conditions that you specifically requested be included in the bid. The project is scheduled for winter construction, so there's no doubt this will be required. You work up your own estimate (or check the other masonry bids to verify the amount they've included in their quotes for winter weather), do the math, and determine this to be an $8,000 line item shortfall. That's no small amount!

You call the "low" bidder back to relay your findings, but he just shrugs it off and says, *"I thought you guys were picking up that cost."* You take a deep breath, control your desire to reach through the phone and grab him by the throat, and just consider yourself lucky you caught the discrepancy before it caught you. You turn back to the other proposals and discover that the third highest proposal was actually the low bid, after you take all of the general requirement provisions into account. This could have been a disaster! It's no fun to be three months into a job, and have the mason come up to you and demand $8,000 for heat and enclosures. At that point, the importance of general requirements becomes abundantly clear!

So let's take a look at some common general requirements, what they are, and what the estimator needs to know to address them correctly. You can break them down into two categories: *administrative* and *field* general requirements.

Keep in mind, as we go along, that the breakdowns we're using for our example in this chapter are for descriptive purposes only. What constitutes *administrative* or *field* general requirements will vary from one builder to another, depending on their circumstance and point of view. It's OK if you want to consider one of our field general requirements in your administrative costs. What's important is that you understand where you want each item to go and why. It's even more important to be consistent about your system so that you (and everyone else in your office) know where these items belong. In short, if you think dumpsters belong in administrative general requirements, put them there. But once you do, *then you have to keep them there!*

Administrative General Requirements

Administrative general requirements are those general requirements that tend to be associated with the office, and aren't directly related to the physical construction of a project. Figure 5-2 shows a listing of some of my typical administrative general requirements.

Architectural and Engineering Fees

Unless you're into a design-build or other semi-negotiated deal, the architectural and engineering cost is generally handled directly between the owner and the design firm. There are, however, situations where you, as the CGC, may be responsible for some additional architectural or engineering liability. Items like engineered truss certification, special beams or structural elements, and mechanical and electrical detailing are just a few to watch for.

Special Construction Considerations

Special construction considerations include items such as:

- Loss of productivity due to working height (first-floor work is easier than the fortieth-floor work)
- Weather conditions (particularly winter)
- Commuting long distances to the site
- Special hoist or crane conditions due to odd terrain or the lack of a staging area
- Poor labor or subcontractor availability
- Lack of on-site material storage
- Limited working or framing space available on the job
- Inexperience, either in your own labor force or your subcontractors'

ABC Contractors

1234 Hammertacker Way, Lienwaiver, WI 53555 • Office (608) 555-1234, Fax (608) 555-1235

Administrative Project General Requirements Worksheet

Code	Description	Qty	Unit	Labor $/unit	Labor total	Material $/unit	Material total	Equipment $/unit	Equipment $/total	Line item total
1.51	Special/additional pre-engineered truss engineering or certification									
1.52	Special/additional structural engineering or certification									
1.53	Special/additional mechanical/electrical engineering certification									
1.54	Special/additional engineering/certification (other)									
1.55	Allowance for additional commuting distances to site by workers									
1.56	Labor and subcontractor availability									
1.57	Material storage facilities & staging area									
1.58	Allowance for limited working or framing space available on job									
1.59	Cost to cover schedule acceleration or liquidated damages									
1.60	Special insurances and bonding (not covered in base proposal)									
1.61	Administrative time for special permit/zoning procedures (meetings, etc.)									
1.62	Special permitting/zoning fees (not covered in base proposal)									
1.63	Allowance for new/inexperience labor (or new work application)									
1.64	Allowance for abnormal mobilization and/or employee lodging									
1.65	Special contingencies (i.e. difficult architect or owner, bad past experience, etc.)									
1.66	Construction project meetings – actual time spent									
1.67	Construction project meetings – preparation of minutes, agendas, etc.									
1.68	Preparation and processing of change orders/requests									
1.69	Preparation and processing of project schedules/scheduling revisions									
1.70	Preparation of submittals, samples, shop drawings & mock-ups									
1.71	Project closeout – preparing operations & maintenance manuals									
1.72	Project closeout – other (warranties, affidavits, closeout documents, etc.)									
1.73	Special accounting/records procedures									
1.74	Special photographs/promotional marketing									
1.75	One-time worker trade certifications, dues, etc.									
1.76	Adjustments for geographical or regional differences (indexes)									
1.77	Adjustments for time/cost indexes (for in-house labor data)									
1.78	Other									
1.79	Other									
1.80	Other									
	Totals									

File: ADMINREQ

Figure 5-2

Administrative project general requirements worksheet

And there are many more items that fall into this category. Beware! These kinds of problems are the slippery ones — they can be easily forgotten or underestimated. Unfortunately, they can also swell into big dollars by the end of a months-long project, eating away at your profits.

The best way to address these items is to step back, after you've considered each individual area of work, and look at the project as a whole. Think about what has to be done that's outside of the areas that you've already considered. Then assign labor, material and equipment costs that you think are adequate to complete that work, based on your experience. These are items that, unlike contingencies, you know in advance you'll need to cover in your bid.

If you're not comfortable with this type of big-picture visualization technique, there are excellent construction estimating books on the market that offer help with work items like these. Many offer labor productivity ratios by trade that you can use as a guide when trying to determine how a worker's production will be affected by outside circumstances. Estimating manuals can give you a place to start until you've accumulated your own in-house database of special labor, material, equipment and productivity costs.

Schedule Acceleration or Liquidated Damages

You can add this cost into your estimate to cover projected overtime hours, temporary motel lodging for your workers (to avoid mobilization time and expense), or anticipated liquidated damages which you'll have to cover due to an unrealistic or ill-conceived project schedule set forth in the bid documents. Unfortunately, this does happen. Let's say, for example, the architect has published a project schedule that you genuinely consider to be unrealistic — about a month short. Furthermore, the contract calls for liquidated damages to the tune of $200 a day for any delay in the schedule. You might want to add $6,000 to your proposal to cover those costs!

Contingencies

A contingency is generally an agreed-upon, fixed amount set by the owner and architect to cover unforeseen expenses. Even though they've already made a provision for this expense, you may be asked to supply your own contingency amount. Or maybe you just want to have your own figured in. This "slush fund" most often covers unforeseen (yet expected in some way) costs that occur after the project has gotten underway.

I've found that these type of surprises come up most often during the excavation process, but it can be different for each contractor. I've run into problems such as an unusually high water table that requires thousands of dollars in dewatering, or having to deal with old abandoned subterranean water lines or utilities that were forgotten by the city engineering department. You know that such unexpected expenses can occur, you just don't know what form they'll take or how much the financial damage will be. That's why you need a contingency.

Project Meetings, Change Requests and Scheduling

Project meetings, change requests and scheduling always require office time — and sometimes a lot of it! You need to figure that cost into your estimate, too. Your time is money. Job progress meetings involve agendas, recording the minutes of the meeting, and then often working out revised project schedules. And the office time spent processing change requests can be absolutely *enormous* — especially if you're working with an owner and architect who didn't do their homework or rushed the project out for bid before all of the details were defined.

Insurance

There are numerous types of insurance you have to carry in this business. Builders' risk, all-risk, floaters, public liability, workers' compensation, employer liability, and more. Insurance is a necessary evil in our industry. I include builders' risk insurance and any other special policies required for a specific job in my estimates, and lump the other insurance costs in with my administrative overhead. Some general contractors may break down a percentage of all their insurance costs and add them into each estimate as a direct cost — it depends on their general business philosophy. There's nothing wrong with including them in the body of your estimate rather than as part of your overhead and profit. Just make sure you include them someplace.

"Whether you're a subcontractor or a general contractor, the submittal process will cost you money."

Bonding

Bonds (purchased through bonding or surety companies) are normally the responsibility of the general or prime contractor. In construction, the most common kinds of bonds are the *bid bond*, which guarantees that you'll honor your written bid, and the *performance and material payment bond*, guaranteeing performance of your work and payment of your suppliers. You need to account for bond fees in your quotation.

Permits

It can be particularly hard to pin down the cost of permits. When you call downtown, the inspector, not having seen the plans, may not be able to commit to any definite costs. So filling in this line item may require using your experience with past jobs. Be careful that you don't duplicate here. For example, does the plumber already have the sewer connection permit?

Submittals, Samples, Shop Drawings and Mock-ups

Whether you're a subcontractor or a general contractor, the submittal process will cost you money. The submittal process (which we'll cover in detail later in the book) is simply a check and balance system set up by the architectural community to ensure that the products and services specified in the plans and specs make it into the construction. The subs and suppliers have to provide the general contractor

with samples which the CGC can forward to the architect and owner before construction of that part of the work begins. For the sub and supplier, this means obtaining product samples from the manufacturers or putting together a collection of Sweet's® literature and adding some minor customization to turn it into a formal submittal package for the CGC. But on some projects (depending on the size and scope of the project, the architect/engineer, or your specialization), you may be asked to provide more detailed submittals. You may need to supply shop drawings that detail the manufacturer's required installation methods, and mock-ups (scale models of the proposed construction) which can be quite costly. The cost of putting together the proposal package, as well as its presentation if that's part of the bid requirements, has to be accounted for when bidding the project.

Taxes

When it comes to the tax line item on your estimate (and on your general requirements worksheet), the taxes you'll most often have to include are sales and use taxes. The responsibility for paying these taxes on a construction project is normally spelled out in CSI (Construction Specification Institute) Division 1 of the specification manual. Public projects are often tax-exempt, but private projects may be taxable on a state, county, or city (or other specialized area) level, so it's important to have them covered in your bid. These are generally the only taxes covered as job line items. Social security, unemployment compensation, and other employer-related taxes are usually handled as administrative overhead and figured into your final overhead and profit percentage.

Project Closeout, Warranties, and Operations and Maintenance Manuals

There's often a considerable amount of office time required for closing out the job. The CGC is responsible for writing and/or gathering and distributing owner-required warranty letters, operations and maintenance manuals, and other closeout documents, such as the substantial completion notice and affidavit of liens. These are also a cost of the job and must be included in the proposal.

Accounting and Pay Requests

Don't forget that considerable time and effort go into administration and record keeping on each project. It would be a mistake to assume that every job requires the same amount of administrative and accounting time. On some jobs, these items can be an enormous profit-eating liability, particularly when working with government agencies, local utilities, or if you're in a partnering arrangement that includes sharing all accounting. The best thing to do is to establish a benchmark amount of office time to be allotted for accounting and administration, based on your typical job. Then, when you project these costs to run higher, the estimator can add those extra costs to that base amount. The more administrative hierarchy and red tape you have to go through to get a job done, the higher the costs will be.

And There's More . . .

Every job is a little different and may have a different set of general requirements to deal with. There may be photos, special filing and/or accounting procedures, trade certifications, or any number of items that you don't run into on every job. What, when, and how to bid these items is mostly a result of experience, plus a careful reading of the specification manual for the project. Identifying and accounting for all general requirement line items is part of good construction management. And so far we've only discussed the office!

Field General Requirements

Field general requirements are the more "nuts-and-bolts" type of expenses that are usually associated with the job site. Let's look at those general requirement items that can sneak up on you out there.

Supervision

If you're a small- to medium-size commercial construction contractor, this may be your largest general requirement line item. It's not uncommon for a superintendent, who may also be one of your lead workers (depending on the size of the job), to spend half, three-quarters, or even all of his time on pure supervision. The larger the job, the more time is required for trade coordination, public relations, settling problems, and other peripheral tasks.

Supervision costs money — the exact same money that's paid to carpenters, masons and electricians. Treat it exactly like a line-item labor cost. If, because of the nature of the project or because it's required in the specs, you feel that 90 percent of your superintendent's working hours will be spent supervising, make sure that time is reflected in your bid.

Another way to do this is to add a separate line item to your actual bid summary. Call it job supervision. That way, you avoid the temptation to just "add a little" to, say, your carpentry line item to cover carpentry supervision. Don't confuse your labor take-off with anything except actual labor hours.

Sit back and map out the job. Invite your superintendent to join in if you think his input would be valuable. Come up with a realistic forecast of hands-off, true supervision manhours. Then multiply the result by your billing rate for supervision, such as $42 per hour, and add that amount in the labor column of your summary sheet. As the jobs get larger and more complicated, you may need many more full-time field personnel working as supervisors, project engineers and safety directors, as well as a site-trailer clerical staff, and more. Each of these people working in the field will be a cost of the job.

Waste and Trash Disposal

I don't know about other contractors, but waste and trash disposal seem to nip me in the pocket a lot. For some reason, even when I think I've got it covered, I come up short in figuring my job costs for dumpsters and trash disposal.

One thing you can try, as a prime contractor, is to ask the subcontractors who use the majority of the dumpster space to help offset the cost of waste disposal. In fact, many CGCs require this. If you're a subcontractor, you may need to include this in your quotation. If the job is beyond dumpsters and you have to truck waste offsite, try to obtain a lump-sum quote for the cost before you finish your bid.

Temporary Construction

There can be a number of reasons for you to have temporary construction on the job site. You may be required to protect adjacent property with walls, provide temporary fencing to keep spectators off the site, or you may have to provide temporary offices for the customer's employees while you're renovating a wing of a building. You may also need temporary structures for storage and for dust or noise control, just to name a few.

Site Personnel Facilities (Trailers)

On most large projects, you'll normally need a job trailer. That's a cost all by itself — but there's more. You have to also consider the cost of transporting the trailer to the job site (and away when the job is done), and providing a level area for the trailer to rest on. There are temporary utilities to be brought into the trailer: phone, fax, heat, lights, drinking water, and portable toilets. And finally, you need to stock provisions for first aid and put up a bulletin board where you can place required job, health and safety postings. All of these are job costs.

Temporary Job Utilities

And while we're discussing temporary utilities, are you required by the specifications to pay for the power you use on the job, apart from the job trailer? Will you need to provide a generator and fuel or temporary hook-ups for service? What about water for the masons? Lighting for workers?

Field Engineering (Survey and Layout)

You may have received a quote from a surveying company for the initial building or project layout. However, there are many times where it's necessary to re-establish grades or benchmarks as the job progresses. These could result in additional charges, so don't forget to check into the possibility.

Project Signage and Temporary Barricades/Traffic Control

Sometimes construction projects require traffic or pedestrian control, especially road work or any construction that's close to pedestrian traffic areas. Barricades, temporary traffic signals or even a traffic-control duty person may be required for public safety. You may also need to get permits and permissions to close off part or all of a thoroughfare. The logistics of controlling traffic around your construction site can be complex enough to require you to hire a specialty subcontractor just to handle traffic control. This can cost a lot of money — and it should definitely be included in your job bid.

Winter Weather Conditions and Snow Removal

Winter weather can cause big added expenses. We're not just talking about shoveling and plowing snow — which is in itself a time-consuming and expensive activity. There are also enclosures to build, the temporary heat needed to warm those enclosures, and of course, the tankloads of fuel required to provide that warmth.

Mobilization, Trucking, Vehicle Expense and Special Freight Charges

If something or someone has to get from point A to point B, it's going to cost you something. You'll also need fuel for site machinery, generators, and more.

Small Tools and Special Equipment Rental

The small tools category covers buying or replacing small tools needed for the job (drills that burn up, etc.) and their blades, bits, bearings, and frayed cords. And what about the occasional rented partner saw, power auger, or pump? Don't forget to include the cost of all special equipment that you rent for a job.

And speaking of special equipment and pumps, *dewatering* a site can range from renting a few pumps and having someone on hand to watch over them, to providing and maintaining full-scale deep wells and hoses and digging a ditch to run off the water. It can run into thousands of dollars. Be sure to read those soils reports carefully!

Temporary Access Roads or Site Maintenance

Access roads can be expensive to build, but they're generally easy to foresee during the bid period. Not so obvious, however, is the maintenance of the roads and the immediate site access. You may need to periodically bring in replacement rock and even out the road surface — especially if it rains. And don't forget, if it's temporary, you'll need to *remove* it all at the end of the project.

Safety Programs

Today, safety is strictly and consistently enforced on the job site — and for good reason. Obviously, there's nothing more important than keeping workers and spectators safe on and around the site. However, safety comes at a cost. You may be required to provide safety equipment (safety glasses, clothing and hardhats), fall protection (netting, railings, etc.), first-aid equipment and breathing apparatus. HAZCOM (short for Hazard Communications) programs are required by OSHA, and some jobs involve site-specific safety programs, such as the interim life safety programs required of workers on projects at medical institutions. They include hospital emergency procedures, patient evacuation, fire safety, code response procedures, and more.

The HAZCOM programs are part of most contractor safety programs. Contractors must supply information to their workers about what to do in the event of exposure to any potentially harmful products used on the job, and they must make sure the workers are able to read and interpret warning labels and understand MDS (Material Safety Data) sheets supplied by manufacturers.

Scaffolding and Staging

If you've been around construction for very long, you know the value of scaffolding and adequate, efficient staging for your workers. In addition to the cost of the scaffolding and platforms themselves, there's the cost of the material handlers, forklifts, hoists, and cranes that may be required to set up these structures.

Field and Material Testing

You may need to pay for field tests for soil, groundwater, and seismic stability, and a service for testing concrete, asphalt, aggregates, bricks and masonry products, reinforcing steel, and welds.

Project Site and Final Building Cleanup

Depending on your discipline or scope of work, you may be required to provide cleaning services after the job's complete. At the very least, you should ensure that your crews clean up after themselves. Don't forget to include in your bid the man-hours and supplies required to accomplish this.

Summary

As complete as this general requirements listing appears to be, it still doesn't address *all* of the possible events or items that can run up costs on the job. Your situation may vary considerably. Hopefully, this information will at least provide you with a checklist of sorts and put you on the right track for developing your own general requirements appraisal and costing program. It's important to keep this in mind at all times: If you suspect, even for a moment, that something is going to cost you money . . . it will! Believe me. I know.

Chapter 6

Creating a Winning Estimate

I wish I could begin this chapter by telling you how much more exciting commercial estimating will be than residential estimating — but I can't. In both kinds of construction, estimating is painstakingly methodical work. The most notable difference between the two is that in commercial estimating, the numbers are usually larger and the pitfalls deeper.

But commercial estimating does have its rewards. Winning the bid for the new city hall project in your hometown generally offers more satisfaction and recognition than winning the bid for another ranch house. How many ground-breaking ceremonies, gold shovels, and pictures with the mayor are there when you begin construction on a private home? When you get your name in the newspaper for being awarded a big commercial project, people remember you and they become interested in your projects — and in you.

Of course, that happens only when and if you get the big jobs. As we mentioned earlier, you won't win every project you estimate; maybe not even one in every ten. So the satisfaction you get from the ones you do win may have to last you awhile. But don't let that get you down. You can find satisfaction (and small, steady doses of self-esteem) from the estimating process itself. No, I'm not joking. If you can take a set of plans and specs for a complex building project and create a cost breakdown of every part — materials, labor, extras, overhead and profit — and come up with a total that's very close to what it'll actually cost, that's quite a feat. Not a lot of people can do that. What would that disparaging grade school math teacher say now? I find I get a lot of satisfaction from creating an estimate, and I've developed some useful tricks and techniques along the road. We'll cover these, and set you on the path to becoming about as good at estimating commercial work as the best in the trade. And that's me! I'm proud of my skill, and so should you be.

Commercial Construction Estimating

Commercial construction estimates can be extremely complex, often involving hundreds of line items and facts to be checked, rechecked and verified. These estimates demand time and attention throughout the process. However, you're limited to a specified bid period, so you need to focus clearly on one line item at a time, address it to a reasonable level of certainty, and then move on to the next item.

This "moving on" part is sometimes difficult for estimators. Like engineers, most of us are obsessive about detail and we can spend hours dwelling on each detail in front of us. But in the real world of competitive estimating, there's normally no time (or need) to obsess over one particular line item. If you address each component with reasonable and responsible attention, then you've done your job. The estimate will come together and meld into a complete, comprehensive and competitive body of work.

True, there *are* occasions when being a little obsessive might be necessary. Let me give you an example. A while back, I bid on a small building for a local wastewater facility. The sole purpose of the building was to house a centrifuge for the treatment plant. The cost of buying and installing the centrifuge was part of the contract, so I included it in my bid. I prepared my estimate as always, line item by line item, and then did the final tally. The cost for the building itself, including accessories and site work, came to $72,000. However, my total bid was $432,000!

Why such a difference? When I got the quote for the centrifuge, it came in well over $300,000. Obviously, when one line item can make such a monumental difference in the outcome of a proposal, it's definitely OK to obsess over that one line item! You can bet that I checked, rechecked and verified the quote for that centrifuge until I was absolutely sure it was right! This was one cost I definitely didn't want to absorb out of my profit.

The Estimating Process

Let's look at the estimating processes itself. The examples in this chapter are based on the competitive bid. I've chosen this bid process because in commercial work the competitive bid is the most common one you'll encounter, at least in the beginning. Although we're dealing with commercial estimating here, the same rules and mechanics also apply to residential estimating.

We can break the estimating process down into six steps:

1. Setup and organization
2. Bid invitations and information distribution
3. In-house estimates and take-offs
4. Gathering sub and supplier quotations
5. Data compilation/spreadsheet summary
6. Presenting the results (the Bid Form)

Now, let's look at each of these steps in greater detail.

Setup and Organization

You've all heard the adage, *the most important part of a building is the foundation*. The same holds true for commercial construction estimating. A solid, well-conceived foundation sets the pace for all work that follows. That's why it's important for you to pay particular attention to the beginning of your estimate to assure it's on sound footing. Thoroughness and attention to detail early on will pay dividends down the road.

Of course, this thoroughness takes time. Don't make the excuse that you're too busy right now and you'll make it up later. Being busy is never an excuse for rushing or ignoring elements of an estimate. Anyway, you'll seldom have more time later than you do now. Discipline is the key to good estimating. Don't skip steps! Taking shortcuts will only lead to missing line items, which leads to work unaccounted for, which leads to *lost profit*.

The Lost Art of Reading

A crucial and often overlooked element of commercial estimating is taking the time to *actually read* the plans and specifications. Now, you're probably saying, *"Isn't that obvious?"* Unfortunately, it's not as obvious as you may think. There are too many estimators who don't put enough time and effort into really understanding the full scope of the project they're bidding. They skip over that time-waster and go right into the estimate.

Perhaps that's understandable. Commercial construction is a very fast-paced, pressure-packed business. Chaos is common and deadlines are perpetually looming before us. You have to put out the fires (problems) in front of you *right now* before you can get to the things that are coming up. It's easy to see how an estimator, buried in current responsibilities, could too quickly glance over the plans and specs and assume he understands them because he's *"seen jobs like this one before."*

It's OK to be sympathetic — just don't *be* one of those estimators. You won't win. It's only a matter of time before mistakes happen. Remember, you're in commercial construction now! You're not selling hamburgers! The product can be very large and very costly, and mistakes can translate into big dollars. Just one can wipe you out! Spending an undisturbed hour or two absorbing the overall intent and scope of the project seems like cheap insurance when you consider the potential impact an error can have on your professional and personal well-being.

If you're the type who works outside the office, you may want to take the plans and specs home in the evening or over a weekend for some undisturbed study time. Yes, I know this cuts into personal time, but the benefits are:

1. You'll be less stressed knowing you're familiar with the job.

2. You'll gain a competitive edge over the other bidders.

3. You'll be a better estimator and therefore up for more opportunities in the future.

The Specification Manual

After absorbing the overall scope of the project, get out the specification manual, if one accompanied the plans, and look over the Table of Contents. It'll give you a feel for the various trades that you'll have to contact for sub and supplier bids.

After checking the Table of Contents, continue on and look over the general administrative information about the project. This is usually found under *Invitation to bid*, which follows the Table of Contents and makes up the first page of the actual text of Division I, General Requirements. This page gives details about the project, such as:

▌ The proper name of the project

▌ The name(s) of the owner(s)

▌ The bid due date

▌ The location where the bids will be received

▌ A brief summary of the project itself

▌ Other information that the owner or architect thinks pertinent to the project bid

Now look at the rest of *Division 1, General Requirements*. Specification manuals for the private sector are often organized in CSI (Construction

CSI Divisions	Description
1	General Requirements
2	Site Work
3	Concrete
4	Masonry
5	Metals
6	Wood & Plastics
7	Thermal & Moisture Protection
8	Doors & Windows
9	Finishes
10	Specialties
11	Equipment
12	Furnishings
13	Special Construction
14	Conveying Systems
15	Mechanical
16	Electrical

(Handwritten annotations: "→ INDIRECT COSTS" pointing to Division 1; "DIRECT COSTS" bracketing Divisions 2–16)

Figure 6-1
CSI divisions found in commercial specification manuals

Specification Institute) format, that is, divided into sixteen construction divisions, beginning with *General Requirements*.

Figure 6-1 shows the headings for the CSI divisions normally found in commercial specification manuals.

CSI does have additional, more specialized divisions, but they don't come up that often in everyday commercial construction. When they do, simply apply the same methods and care as you normally would and include the cost in your final estimate summary. Each CSI heading represents a hierarchy of *divisions*, *subdivisions*, and *even subsubdivisions*. If you'd like more information on CSI, or would like to learn how to become a member, you can contact them at:

Construction Specifications Institute
601 Madison Street
Alexandria, VA 22314
Phone 1-800-689-2900
http://www.csinet.org

The division that you focus on as an estimator is determined by the type of work your company performs and bids. For example, a plumbing contractor

DIVISON 1 - INDIRECT COST

will focus most of his time on CSI Division 15, and a carpentry subcontractor on Division 6. General contractors and commercial general contractors have to know something about all the divisions — and give special attention to those divisions in which they have crews who perform in-house work.

Division 1 — General Requirements — No matter who you are, you're also responsible for reading and complying with all the information in Division 1, General Requirements. This division outlines the responsibilities of each party (owner, architect/engineer, general contractor, subcontractors, and suppliers) in the administration of the project. General Requirements cover the information that isn't associated with the actual "hands-on" construction of the project. This division usually contains items such as:

▐ Bid package and bid documents

▐ Insurance requirements

▐ Bonding requirements and forms

▐ Prevailing wage rate and overtime requirements

▐ Job supervision requirements (Will you have to pay for a full-time superintendent?)

▐ Project administrative requirements such as scheduling, job meetings, or trailers

▐ Warranty, Operations and Maintenance Manual, and as-built requirements

▐ Special bid considerations such as tax status, allowances, testing, or cut and patch

▐ Change order and contract modification parameters

▐ Any other instructions or information that the architect, engineers, and owners want the bid contractors to know about how they wish the project to be handled administratively

Read Division 1 as completely as possible. I say as completely as possible because some of these specs are very lengthy. You simply may not be able to give this division a truly careful read and still be able to cover all of Divisions 2 through 16, given the length of a normal workday. You'll probably need to prioritize your need for information and pick out the areas to read carefully. The ability to do this comes with "bidding maturity," and gets easier with the number of jobs you estimate. As you bid commercial jobs, you'll learn to differentiate between *boilerplate* (repetitive and common architectural jargon) and information that's specific to the project you're bidding.

The requirements and obligations in Division 1 all cost you money in some way, so they need to be accounted for in your bid. When compiling an estimate, these general requirement items are commonly referred to as *indirect costs* (or sometimes just *indirects*). Once you've completed a number of estimates, you'll have a clearer idea of what these indirect costs include. You'll learn to recognize out-of-the-ordinary items and supplementary or special notes that could increase your normal general requirement costs. Items like a full-time superintendent, liquidated damages, night watchmen and security requirements, and in-house training programs after the job's complete are a few of the abnormalities that should be a red flag for any estimator.

Direct Costs vs. Indirect Costs — *Direct costs* are generally those costs associated with CSI Divisions 2 through 16. They're usually part of the "hands-on" construction costs. Separating indirect from direct costs can sometimes be difficult, but it's more important to be consistent about where you place items in a bid than to be concerned with what truly constitutes an *indirect* item. An example would be the general "cut and patch" responsibilities commonly found in Division 1. Since it's part of "hands-on" construction, should it be a direct cost or an indirect cost? That's a good question. However, the answer isn't important. Just decide where you want to put it, and then put it there consistently on every bid.

The "Line" — There's one other thing you've got to consider when weighing directs and indirects. That's whether or not the item will eventually be *marked up* (an additional percentage added on for your profit and overhead) as part of the final quotation. The "line" is merely a division between those items that the contractor thinks should be marked up (above the line) and those items that shouldn't (below the line).

Determining which items should be marked up and which shouldn't, is again, purely individual. One contractor may choose to mark up everything, and others only direct costs. Competition will probably dictate what you may or may not mark up. It's likely that you'll only mark up your direct costs in order to become or stay competitive.

Here's a word of warning: It's very easy to focus your energies on the direct line items in Divisions 2 through 16 and pay less attention to the indirect items. After all, you get information for your direct costs from sub and supplier quotations. Direct cost items are a lot easier to calculate and transfer to an estimate summary than the more evasive indirects. But never, ever underestimate the effect of indirect costs on your bottom line. Telephone bills use up the same kind of money as carpentry labor. And, it's very often indirect cost overruns that cause a potentially profitable job to become a "Well, at least we kept our guys working" kind of job.

Profit — More often than not, the profit line item — which should be the result of a carefully-conceived plan based on your company's financial requirements — is more a product of your local market conditions, aggressiveness, intuition, and need at any particular time. There are, however, some generalizations about profit that I've come to over the years:

▌ First, as the dollar volume for the project goes down, the profit percentage or markup generally goes up. This is clearly illustrated in Figure 6-2. The markup percentage becomes an especially important consideration with very small jobs — say under $5,000. The logic is simple. It just doesn't make much sense (unless you have other motivations such as developing new or on-going relationships) to get 10 percent on $2,000 (or $200), when it costs $42 an hour just to leave the office. Most contracting firms have a minimum profit that they need to generate from a job before they even consider taking the project. For example, I expect a minimum profit of $5,000. If I don't think we'll make at least that much on a job, we don't pursue it. This minimum varies from company to company. Usually, the larger the company's volume, the higher the minimum profit they expect to make on each job.

Actual cost of project ($)	Markup (%)	Markup ($)
2,000,000	4	80,000
1,500,000	5	67,500
1,000,000	6	60,000
750,000	7	52,500
500,000	9	45,000
250,000	11	27,500
100,000	14	14,000
50,000	18	9,000
25,000	25	6,250
10,000	50	5,000

Figure 6-2
Markup percentage increases as project cost decreases

▌ Your percentage of profit should be adequate to cover the growth plans that you've laid out for your business. Or better put, you need to make enough net profit (after taxes and overhead) to provide for future activities, like that larger office you want to move into or the capital expenditure for the new excavating equipment you've ordered. This forecast is part of the basic business plan that every owner needs to map out when they launch their business, and then periodically reevaluate to check their progress and make adjustments.

▌ You'll likely have an occasional job that just breaks even. But if all other aspects of your business are positive and in order, you should still have a healthy profit at the end of the year. If you don't, then you should seriously consider an increase in your markup.

Computers and Construction Estimating

Having succumbed to "computer-geekdom" ages ago, I do all of my estimating, data compilation, and word processing on the computer. However, being comfortable with computers doesn't automatically make you comfortable with individual software programs, and vice versa. Over the years, I've run the gamut of software programs, from full-blown construction estimating systems that incorporate a digitizer table and cost databases, to the not-so-full-blown systems that you can download off the Internet. And many of these systems were wonderfully crafted and marvelously engineered. But somehow, I never really found one that precisely suited me. I think there were several reasons for this.

1. The systems were too bulky and cumbersome for everyday use; that is, it wasn't worth the effort to "crank up the machine" for small bids.

2. They were too conservative in their pricing, and customizing prices within the programs wasn't a quick process.

3. The manhours were unrealistic.

4. They were just too time-consuming to learn and use. First there was the initial learning period, then retraining for revised editions, then inputting the updates, customizing and creating assemblies, and after all that, they never seemed to work as promised.

There are many, many estimators who'll say I'm nuts and praise the system that they use — and that's great! For my part, I created and customized my own estimating spreadsheets in *Excel*. These work great for me. But I know that not everyone is as into computers as I am, and may not feel up to the task of writing their own spreadsheets. Or they may have found a program that does the job and is easy to use.

My recommendation is for you to search out a program or method that works best for you and your work. When you find one that's right, or that comes close, customize it to suit your needs, and then never stop fine-tuning it. Stay open-minded. There are always new ideas in the building industry, and markets and attitudes change almost daily. The intelligent and progressive contractor isn't afraid to change along with them.

National Estimator — Craftsman Book Company, publisher of this book, offers its own estimating software, and gives it away with each of its annual cost books. The program, called *National Estimator*, is about the easiest to use of any of the programs out there. It works like a book. You just turn to any page, or use the electronic index to locate the costs you need. Then a mouse-click copies and pastes cost estimates to your electronic estimating form. It extends and totals the costs automatically. The CD in the back of this book has the *National Estimator* program and 300 pages of labor and material costs for commercial construction. To give this program a try, just insert the CD in your computer.

But even with all the choices available out there, I found myself gravitating back to, and almost exclusively using, my own custom-made spreadsheets. Once I became familiar with the functions and benefits of the common spreadsheet software programs, I grew more enthusiastic about creating spreadsheets for virtually every aspect of the estimating process. It was fun, and the computer skills I picked up along the way didn't hurt my resumé any.

And there was another benefit of creating and customizing my own estimating tools. I simply felt like I was more in control of my estimate than I did with the other methods. My spreadsheets were familiar and comfortable to work with. Once you reach that stage, then becoming competitive is only a matter of time. An estimator who's in control of his estimate is more willing to aggressively tackle a bid and take intelligent, calculated risks. It's a real confidence-builder.

The Estimate Summary Spreadsheet

Figure 6-3 is an example of a typical commercial construction summary spreadsheet; it's the kind I use. If you want a more complete listing, Figure 6-4 provides a very comprehensive example based on the 16 CSI divisions. You could use this one as a prototype for generating your own estimate summary — just list as many of the items as you commonly use. You can pattern an estimate summary spreadsheet after either one of these, or make up one of your own. What's important is that you feel comfortable with how your information is gathered, organized, calculated, and presented.

SAMPLE PROJECT
SERVICE CENTER
YOURTOWN, WISCONSIN
ARCH: ABC ASSOCIATES

11:20 AM
7/10/00

BUDGET: $1,400,000
BID DUE: TUESDAY, 7/14/00, 2 PM
WHERE: ARCHITECT'S OFFICE

CSI	Item	Who	Labor	Material	Sub/Equip	Totals
1000	GENERAL REQUIREMENTS (NOT INCL. SUPERVISION)	GENERAL BLDR	18,334	15,000		33,334
						0
2110	SITE CLEARING, EARTHWORK, EXCAV & BF, EROS CONTROL	SITEGUY, INC.			36,190	36,190
2511	ASPHALT PAVING, STRIPING, & GRAVEL BASE	SITEGUY, INC.			22,493	22,493
2670	WELL WATER SYSTEM	WELL'NGOOD			15,000	15,000
2711	FOUNDATION DRAINAGE SYSTEM (DRAIN TILE)	PLUMBER			4,200	4,200
2731	MOUND SEPTIC SYSTEM	SITEGUY, INC.			12,950	12,950
2900	LANDSCAPING & OUTCROP STONE @ RETAING WALL	LANDSCAPE, INC.			12,132	12,132
						0
3300	CIP CONCRETE - WALLS & FOOTINGS	GENERAL BLDR	34,371	18,289		52,660
3300	CIP CONCRETE - BUILDING FLATWORK & CONC TOPPING	GENERAL BLDR	33,647	11,065		44,712
3300	CIP CONCRETE - SITE CONCRETE & SIDEWALKS	GENERAL BLDR	4,667	1,906		6,573
3300	CIP CONCRETE - RIGID INSUL & ACCESSORIES	GENERAL BLDR	1,580	3,216		4,796
3300	CIP CONCRETE - REINFORCEMENT	GENERAL BLDR	4,500	4,510		9,010
3410	STRUCT PRECAST CONC - PLANTCAST (INCL CAULKING)	GENERAL BLDR			37,134	37,134
						0
4200	UNIT MASONRY & (4711) MANUFACTURED STONE	BRICKENBRACK			192,215	192,215
						0
5120	STRUCTURAL STEEL & METAL FABRICATIONS	METAL/ABC ERECT		20,620	4,400	25,020
						0
6100	ROUGH & MISC. CARPENTRY (INCL PINE PANELING)	GEN/LUMB SUPP	39,960	47,800		87,760
6185	STRUCTURAL GLU-LAM TIMBER TRUSSES	GEN/TRUSS SUPP	1,632	36,447	1,220	39,299
6192	PREENGINEERED MET PLATE WOOD TRUSSES	GEN/TRUSS SUPP	2,176	12,906	360	15,442
6200	FINISH CARPENTRY - RUNNING TRIM (PLAIN SAWN OAK)	GEN/LUMB SUPP	5,440	17,980		23,420
6402	INT ARCH MILLWK - COUNTERTOPS, SHELVES, S & P, CASEWK	CASEWORK, INC.	ABV	ABV		0
						0
7120	FLUID-APPLIED WATERPROOFING	TAR-NATION			4,608	4,608
7210	BUILDING INSULATION	WARM&GO			8,920	8,920
7311	ASPHALT SHINGLES & METAL EDGING	SHINGLE GUY			17,990	17,990
7400	VINYL SIDING, VINYL SOFFIT, & ACCESSORIES	SHINGLE GUY	6,440	7,311		13,751
7600	FLASHING AND SHEET METAL	METALMAN, INC.			ABV	0
7720	ROOF ACCESSORIES	METALMAN, INC.			555	555
7901	JOINT SEALANTS	STOPLEAK CO.	1,920	1,088		3,008
						0
8111	STANDARD STEEL DOORS AND FRAMES / HARDWARE	GEN/DOOR SUPP	10,064	16,326		26,390
8211	FLUSH WOOD DOORS	GEN/DOOR SUPP	ABV	ABV		0
8335	OVERHEAD COIL GRILL	O.H. DOOR CO.			1,637	1,637
8360	SECTIONAL OVERHEAD DOORS	O.H. DOOR CO.			1,042	1,042
8410	ALUMINUM ENTRANCES AND STOREFRONTS & GLAZING	GLASS 'N GO			26,500	26,500
8600	WOOD WINDOWS	GEN/LUMB SUPP	* IN 6100	36,175		36,175
						0
9250	GYPSUM BOARD	WHATAGYP	17,444	11,632		29,076
9300	HARD TILE	QUARRY, INC.			17,780	17,780
9511	ACOUSTICAL PANEL CEILINGS	CEILING SUB			13,929	13,929
9660	RESILIENT FLOORING & CARPETING	FLOORFIRM			14,200	14,200
9900	PAINTING	DROPCLOTH, INC.			29,000	29,000
						0
10155	TOILET COMPARTMENTS	PARTITION, INC.			2,949	2,949
10350	FLAGPOLES	GEN/PATRIOT CO.	544	1,500	650	2,694
10425	SIGNS (25) CAST ALUM LETTERS + INT ADA SIGNAGE	GEN/DIV10GUY	765	1,610		2,375
10522	FIRE EXTINGUISHERS, CABINETS, & ACCESSORIES	GEN/DIV10GUY	425	530		955
10650	OPERABLE PARTITIONS	GEN/DIV10GUY			5,082	5,082
10800	TOILET AND BATH ACCESSORIES	GEN/DIV10GUY	1,122	2,085		3,207
11450	RESIDENTIAL APPLIANCES (* EXHAUST HOOD ONLY)	GEN/DEPT. STORE	* IN 16000	139		139
						0
14245	HYDRAULIC ELEVATOR (W/ PHASE ADDER - INSTALL BY 16)	UPNDOWN, INC.			25,396	25,396
						0
15400	PLUMBING & PIPING	PLUMBER			27,900	27,900
15600	HVAC & GAS PIPING	HEAT 'N COLD		3,500	106,000	109,500
16000	ELECTRICAL & LIGHTING	ZZZTTT, INC.			128,630	128,630
	SUBTOTAL		185,031	271,635	771,062	1,227,728
	OH & PROFIT			6.0%		73,664
	BOND			0.9%		11,050
	SALES TAX			5.50%		13,446
	LIQUIDATED DAMAGES & SPECIAL CONTINGENCY			FIXED		0
	SUPERVISION			FIXED		19,544
		GRAND TOTAL (JOB)				**1,345,431**

Figure 6-3
Estimate summary spreadsheet

ABC Contractors

1234 Hammertacker Way, Lienwaiver, WI 53555 • Office (608) 555-1234, Fax (608) 555-1235

CSI Estimator

Task description	Materials				Labor				Sub bids equip./rent	Task total
	Unit	Amt.	Cost	Total	Unit	Amt.	Cost	Total		
General conditions										
Fees										
Fee – City license										
Fee – Parks & recreation										
Fee – Plan check (general)										
Fee – School district										
Fee – Sewer assessment										
Fee – Miscellaneous										
Fee – Coastal development										
Fee – Dedication										
Fee – Environmental health										
Fee – Front end advances (utilities)										
Fee – Mitigation										
Fee – Parking										
Fee – Safety OSHA										
Fee – Water share rights										
Fee – Transportation corridor										
Fee – Utilities tie-in										
Fee – Water department (water meters)										
Permits										
Permit – Building										
Permit – Demolition										
Permit – Grading										
Permit – Power pole (temporary)										
Permit – Public work										
Permit – Retaining or property line walls										
Permit – Electric										
Permit – Encroachment										
Permit – Fire sprinklers										
Permit – Fireplace										
Permit – Mechanical										
Permit – Plumbing										
Permit – Sewer cap										
Permit – Street use/dumpster/etc.										

Figure 6-4

Estimate summary spreadsheet in CSI format

ABC Contractors

1234 Hammertacker Way, Lienwaiver, WI 53555 • Office (608) 555-1234, Fax (608) 555-1235

CSI Estimator (continued)

Task description	Materials				Labor				Sub bids equip./rent	Task total
	Unit	Amt.	Cost	Total	Unit	Amt.	Cost	Total		
Temporary on site										
Dumpster										
Electric meter & power billing										
Fence & gates										
Job shed or trailer										
Phone/fax line										
Power pole & equipment										
Rental equipment										
Job toilet										
Water billing										
Temporary miscellaneous										
Barricades										
Container storage										
Crane										
Dewatering										
Dust protection										
Progress photos										
Protection, finishes										
Protection, pedestrian (walk with cover)										
Protection, weather										
Security lighting										
Superintendent (full/part time)										
Carpenter										
Finish carpenter										
Semi-skilled labor										
Day labor										
Transportation										
Contractor's insurance										
Builder's risk (fire & theft)										
Performance bond										
Public liability & completed operations										

Page 2 of 12

Figure 6-4 (continued)
Estimate summary spreadsheet in CSI format

ABC Contractors

1234 Hammertacker Way, Lienwaiver, WI 53555 • Office (608) 555-1234, Fax (608) 555-1235

CSI Estimator (continued)

Task description	Materials				Labor				Sub bids equip./rent	Task total
	Unit	Amt.	Cost	Total	Unit	Amt.	Cost	Total		
Miscellaneous conditions										
Architectural drawings										
Engineering consultants										
Bid procedure										
Blueprints & specifications										
Certified inspectors										
Perishable supplies billable to job										
Cleanup final subcontract										
General conditions miscellaneous										
As-built drawings										
General contractor consulting										
Operation & maintenance manuals										
Product warranties packages										
Sitework										
Demolition										
Concrete coring										
Cut, break & remove all concrete indicated										
Demolition, major										
Tree removal										
Asbestos removal										
Shoring costs										
Grading										
Backfill & compact										
Backhoe trench										
Footing soils remove										
Erosion control & drainage										
Grading rough										
Import/export (allowance)										
Utilities to building										
Fire hydrant										
Relocate/remove power pole										
Relocate/remove street light										
Cable TV service										
Electrical service										

Page 3 of 12

Figure 6-4 (continued)

Estimate summary spreadsheet in CSI format

ABC Contractors

1234 Hammertacker Way, Lienwaiver, WI 53555 • Office (608) 555-1234, Fax (608) 555-1235

CSI Estimator (continued)

Task description	Materials				Labor				Sub bids equip./rent	Task total
	Unit	Amt.	Cost	Total	Unit	Amt.	Cost	Total		
Utilities to building (continued)										
Fire sprinkler service										
Gas service										
Telephone service										
Laterals sewer										
Storm drainage										
Street signs										
Asphalt										
Patch										
Paving										
Slurry coat										
Striping & bumper										
Gates & fences										
Fence (chain link, wood, glass)										
Gates (chain link, wood, glass)										
Wrought iron or metal										
Landscaping										
Backflow preventer										
Landscape lighting										
Planters, preformed										
Stepping stones										
Planting & irrigation										
Concrete										
Footings										
Gunite										
Offsite (sidewalk/curb/gutter/driveway)										
Onsite (swale/curb/sidewalk/etc.)										
Piles & caissons										
Precast products										
Retaining wall, concrete (poured in place)										
Masonry										
Fencing or planters										
Retaining walls										
Trash enclosure										

Figure 6-4 (continued)
Estimate summary spreadsheet in CSI format

ABC Contractors

1234 Hammertacker Way, Lienwaiver, WI 53555 • Office (608) 555-1234, Fax (608) 555-1235

CSI Estimator (continued)

Task description	Materials				Labor				Sub bids equip./rent	Task total
	Unit	Amt.	Cost	Total	Unit	Amt.	Cost	Total		
Miscellaneous sitework										
Soils engineering & inspections										
Civil engineering & survey										
Stucco walls/trash area/planter/etc.										
Painting gates/fence/bollards/other										
Bridge										
Docks & boat facilities										
Shoring & bulkheading										
Sitework miscellaneous										
Cap sewer										
Drainage systems										
Soil poisoning or treatment										
Vaults & vault doors										
Concrete										
Foundations (footings & slab)										
Concrete patch work										
Gypcrete, lightweight, etc.										
Lab testing, concrete										
Prestress concrete construction										
Tilt-up construction										
Masonry										
Block foundation walls										
Glass block										
Lab testing, masonry										
Masonry miscellaneous										
Masonry special finishes (cast concrete columns)										
Sandblasting										
Stonework										
Veneers										
Metal										
Metal – ornamental (handrails/railings, etc.)										
Registered inspector, metal										
Structural steel										
Steel miscellaneous										

Figure 6-4 (continued)

Estimate summary spreadsheet in CSI format

ABC Contractors

1234 Hammertacker Way, Lienwaiver, WI 53555 • Office (608) 555-1234, Fax (608) 555-1235

CSI Estimator (continued)

Task description	Materials				Labor				Sub bids equip./rent	Task total
	Unit	Amt.	Cost	Total	Unit	Amt.	Cost	Total		
Metal (continued)										
Metal – equipment screens										
Metal – light gauge frame										
Metal siding										
Metal – shop drawings										
Metal – stairs/spiral/access ladders										
Metal – structural connections										
Structural steel specialties										
Carpentry										
Carpentry, rough & light hardware										
Light framing hardware										
Lumber, glue laminated beams										
Lumber, rough										
Lumber, truss or joint										
Lumber list (send out for builder's list)										
Carpenter rough set exterior doors & windows										
Carpentry finish										
Cabinets, millwork installed										
Architectural millwork										
Fireplace mantel										
Special railings										
Wood base & casing (material)										
Plastic, special fabrications										
Plastic laminates										
Lumber, finish										
Lumber, wood siding										
Wood siding, labor to install										
Storage shelving										
Moisture protection										
Caulking & sealants										
Insulation/air infiltration										
Hot mop (tub & shower base)										
Roof accessories										
Roof patch										
Roofing										
Roofing, metal fabrication										

Page 6 of 12

Figure 6-4 (continued)

Estimate summary spreadsheet in CSI format

ABC Contractors

1234 Hammertacker Way, Lienwaiver, WI 53555 • Office (608) 555-1234, Fax (608) 555-1235

CSI Estimator (continued)

Task description	Materials				Labor				Sub bids equip./rent	Task total
	Unit	Amt.	Cost	Total	Unit	Amt.	Cost	Total		
Moisture protection (continued)										
Rain gutters										
Sheet metal, standard										
Sheet metal, special fabrication										
Skylights										
Waterproofing										
Weatherstripping										
Door, Window, Glass										
Doors										
Coil roll-up door										
Garage door										
Powered operators, garage door openers										
Security, iron type (rolling or overhead)										
Access doors										
Entry doors & frames										
Fire doors & frames (1, 2 hour)										
French doors & frames										
Metal doors & frames										
Specialty doors & frames										
Wood doors & frames										
Glass										
Glass bath enclosure										
Glass screen walls										
Glass special application										
Mirror										
Stained/beveled glass/etched										
Storefront glass										
Wardrobe (mirror, vinyl, other)										
Window glazing										
Windows, aluminum										
Windows & French doors, install										
Windows, wood										

Figure 6-4 (continued)

Estimate summary spreadsheet in CSI format

ABC Contractors

1234 Hammertacker Way, Lienwaiver, WI 53555 • Office (608) 555-1234, Fax (608) 555-1235

CSI Estimator (continued)

Task description	Materials				Labor				Sub bids equip./rent	Task total
	Unit	Amt.	Cost	Total	Unit	Amt.	Cost	Total		
Hardware										
Finish (knobs, latches, closers)										
Installation of hardware (labor)										
Medicine cabinets										
Toilet & bath accessories										
Finish										
Decking										
Elastomeric, mer-kote, dexotex, other										
Hot mop										
Wood decking										
Drywall										
Metal stud framing										
Drywall & metal studs, plus hang, tape & texture										
Drywall – hang, tape & texture										
Drywall – special finish										
Flooring										
Carpet & pads										
Floor preparation (subfloor, float, other)										
Vinyl base										
Vinyl flooring										
Wood flooring										
Painting										
Paint – exterior only										
Paint – interior & exterior										
Paint – interior only										
Paint – special finishes										
Paperhanging										
Lath & plaster										
Plaster – exterior only										
Plaster – interior & exterior										
Plaster – interior only										
Plaster – tile backing										

Figure 6-4 (continued)

Estimate summary spreadsheet in CSI format

ABC Contractors

1234 Hammertacker Way, Lienwaiver, WI 53555 • Office (608) 555-1234, Fax (608) 555-1235

CSI Estimator (continued)

Task description	Materials				Labor				Sub bids equip./rent	Task total
	Unit	Amt.	Cost	Total	Unit	Amt.	Cost	Total		
Tile										
Marble										
Stone										
Terrazzo										
Tile										
Special finish										
Acoustical ceilings										
Acoustical treatments										
Marlite										
Top/simulated marble										
Top/Corian										
Top/unilav (top w/molded sink complete)										
Paneling										
Upholstered treatments										
Specialties										
Access floors and walls										
Awnings										
Chutes (laundry, trash, etc.)										
Directories										
Fireplace, masonry										
Fireplace, prefab metal										
Louvers & vents										
Luminous ceilings (wood frame)										
Pest control										
Postal facilities										
Screens										
Security grilles										
Signage & graphics										
Stairs, construction (wood, metal, hidden)										
Toilet partitions										
Attic access stair										
Shutters, exterior										
Wardrobe & closet specialties										

Figure 6-4 (continued)

Estimate summary spreadsheet in CSI format

ABC Contractors

1234 Hammertacker Way, Lienwaiver, WI 53555 • Office (608) 555-1234, Fax (608) 555-1235

CSI Estimator (continued)

Task description	Materials				Labor				Sub bids equip./rent	Task total
	Unit	Amt.	Cost	Total	Unit	Amt.	Cost	Total		
Equipment										
Appliance (stove/hood/dishwash/disp/micro)										
Appliance (washer & dryer)										
Central vacuum										
Safe or vault										
Equipment miscellaneous										
Furnishings										
Blinds, shades & shutters interior										
Drapery & curtain hardware										
Interior plants & planters										
Moveable partitions										
Furnishings miscellaneous										
Special construction										
Jacuzzi & equipment shed or vault										
Pool or spa decking										
Swimming pool & equipment shed or vault										
Sauna & equipment										
Fountains or waterscape										
Tennis courts										
Wine storage room										
Conveying										
Dumbwaiter										
Elevators										
Wheelchair lift										
Mechanical										
HVAC										
Heating & air conditioning										
Fan special use										
Ducts for all ventilating fans										
Registers, grilles & diffusers										
HVAC shop drawings										
Exhaust fan system										
Refrigeration										

Page 10 of 12

Figure 6-4 (continued)

Estimate summary spreadsheet in CSI format

ABC Contractors

1234 Hammertacker Way, Lienwaiver, WI 53555 • Office (608) 555-1234, Fax (608) 555-1235

CSI Estimator (continued)

Task description	Materials				Labor				Sub bids equip./rent	Task total
	Unit	Amt.	Cost	Total	Unit	Amt.	Cost	Total		
Fire equipment										
Fire sprinkler system										
F.S. shop drawings										
Alarm special application										
Halon system										
Hose cabinet – racks – reels – hose										
Extinguishers – and cabinets										
Plumbing										
General plumbing										
Fixture trim										
Fixtures, Jacuzzi tub										
Fixtures, plumbing										
Fixtures, tub/shower (fiberglass)										
Plumbing shop drawings										
Cesspool/septic tank/pump										
Sump pump/sewer injector										
Wall heaters										
Solar system										
Electrical										
Electric, general										
Exhaust fans, interior (bath/kitchen/laundry)										
Shop drawings										
Lamps & bulbs										
Low voltage – alarm, burglar										
Low voltage – cable TV, prewire										
Low voltage – communications (central)										
Low voltage – emergency lights										
Low voltage – music system										
Low voltage – telephone, prewire										
Low voltage – TV antenna										
Parking lot lighting										
Smoke detectors										
Door bell										

Page 11 of 12

Figure 6-4 (continued)

Estimate summary spreadsheet in CSI format

ABC Contractors

1234 Hammertacker Way, Lienwaiver, WI 53555 • Office (608) 555-1234, Fax (608) 555-1235

CSI Estimator (continued)

Task description	Materials			Labor				Sub bids equip./ent	Task total
	Unit	Amt.	Cost	Total	Unit	Amt.	Cost	Total		
CSI summary										
General conditions										
Sitework										
Concrete										
Masonry										
Metal										
Carpentry										
Moisture protection										
Doors, Windows & Glass										
Finish										
Specialties										
Equipment										
Furnishings										
Special construction										
Conveying										
Mechanical										
Electrical										

Contingency _____ %: _____
Profit _____ %: _____
Overhead _____ %: _____
Total: _____

Page 12 of 12

File: CSIEST

Figure 6-4 (continued)

Estimate summary spreadsheet in CSI format

Earlier, we talked about establishing a foundation for your estimate. Well, there's no more stable ground to rest your foundation on than the summary spreadsheet. Here's how I set up mine (like the one in Figure 6-3):

Step 1 — Pull up the blank prototype copy of your computer spreadsheet (or you can create a template) and save it under a file name for your new project.

Step 2 — Then fill out the top of the page based on information gathered from the front of the specification manual, usually from the Invitation to Bid page. This generally includes the name and location of the project, the bid date, the architect and engineer, the budget (if available), and then the actively current date and time (usually automatically updated by your computer) of the sheet that you're working on. The current time and date functions are available with most spreadsheet software programs. It's an important function, because you can easily print out several versions of your bid and get them confused if the time and date isn't on each.

Step 3 — Next, in the main body of the document, begin listing the job-specific line items using the CSI numbers and descriptions from the specification manual. Since you probably don't know who'll be doing most of the subcontracted or supplier work yet, leave them blank or fill in your company's name for right now.

Step 4 — Now go through the working drawings (blueprints) and enter any additional line items that the CSI format doesn't list or break out as a line item. You need to go through the plans as well, because, unfortunately, it's quite common for the specification manual to miss some items. These will show up when you look at the plans. There are a number of reasons why these items might be missing, but the items (unless otherwise noted) still need to be included on your bid — and listed on your estimate summary.

As the bid process proceeds and you receive new information and addenda, you can add, delete, combine, and otherwise change these line items until you reach your final goal. That's normal. If you're not used to working with spreadsheets, believe me, it will take no time at all for you to become quite familiar with the cut, paste, and copy commands of the software.

Step 5 — Fill in any "below-the-line" items (those you don't want to mark up) such as profit and overhead, architectural and/or engineering fees, bond costs or insurance.

Step 6 — Use the bottom of the page for special notes, calculating square foot costs (to check yourself as you go along), or anything else you feel is relevant to the bid.

Once you've created your summary spreadsheet, it's a relatively straightforward process to enter the results from your own in-house estimates, subcontractor quotations, and supplier proposals in their proper slots. You can enter any cost changes quickly, and the columns are automatically recalculated at the bottom of the page. The benefit of this spreadsheet function becomes very clear as bid time grows closer and proposals and quotations come in fast and furious.

Bid Invitations and Information Distribution

Now, it's time to contact the players. Most construction companies, unless they're very large, do some in-house work. The work not performed by your crews is outsourced to subcontractors, suppliers, and other interests. These are the companies that you need to solicit for bid proposals. Figure 6-5 shows a typical bid log that you can use to track your bid solicitations.

Faxing Bid Requests

This sounds like easy work, but the logistics of "getting the word out" can be quite challenging. I've tried many methods over the years (telephone, postcards, advertising, etc.), but nothing seems to work better or faster than the fax requests that I use now. With the advent of fax software programs, which allow you to fax right off your computer, you can set

ABC Contractors

1234 Hammertacker Way, Lienwaiver, WI 53555 • Office (608) 555-1234, Fax (608) 555-1235

Subcontractor/Supplier Bid Log

Job name: _____

Date: _____

Owner/client: _____

Project number: _____

Architect/engineer: _____

Sheet number: _____ of _____

CSI #	Bidding company	Bid package sent/notified	Bid received	Bid plan and spec?	Addenda received	Notes

File: BIDLOG

Figure 6-5
Bid log

up a database of prospective bidders, create a "bid request" memo, click on those you wish to respond to your bid request, and hit "send." The requests are faxed automatically, and you can even switch to another window application and do other work while they're being processed. It's a fast and efficient way to contact prospective bidders. Figure 6-6 shows a simple fax bid request form that you can duplicate on your word processing program and then use with an accompanying fax program.

Circulating Plans and Specifications

Once the participants are notified, they'll want to see the plans. Everyone in this business knows what a pain in the neck plan distribution is. There are the reproduction costs and clerical time, plus the pure physical bulk of the packages can make it hard to get the information into everyone's hands.

But, there are ways around some of these obstacles. First, there are the plan rooms that we discussed in Chapter 3. They're often conveniently located and offer plans and specifications on local and regional projects for public inspection (although this service may be limited to those who subscribe to the service). Often, you have to do your take-off right there, but some plan rooms will let you check out documents overnight or over a weekend. Many also offer their services on-line, for a fee. This may work for those who have invested in the larger, more expensive computer monitors. On a small monitor screen, however, the plans aren't always as clear as you'd like them to be. Downloading the plans is an option with some services, but there's an additional fee for that.

Other options available to you and your subs and suppliers include:

▌ Having the sub/supplier acquire the documents directly from the architect (this may involve a deposit)

▌ Shrinking and faxing the plans to prospective bidders (requires office and clerical time)

▌ Leaving a master copy at a local copy service and having the sub/supplier pay for their copies

▌ Setting aside a plan room in your own office where prospective bidders can come to do take-offs (may involve increased overhead for office space)

You're really only limited by your imagination and staffing capabilities. Choose whatever works best for you and your situation.

In-House Estimates and Take-Offs

The line items on your summary sheet that aren't covered by bids require your own in-house labor and material estimates. In our office, we work up line item estimates on excavating, concrete, carpentry, and most of the labor for CSI divisions 8.0 and 10.0 in the specification manual. We may include other work as well, if we can handle it with our own crews.

This isn't meant to be a chapter on how to do individual material and labor estimating. You should have a pretty good idea of how to do that by now. But I've included a unit estimate worksheet (Figure 6-7) and an estimate summary sheet (Figure 6-8) which you might find helpful.

If you need more help doing your own take-offs, there are some excellent resources offered by *Craftsman Book Company* and *R.S. Means* that cover construction line item costs, manhours, crew selections, and much more. Both companies offer a variety of guides related to estimating for all fields of the construction industry. There are also guides available through your local library or professional organizations.

Alternates and Voluntary Substitutions

Anyone bidding commercial construction is *obligated to bid the job plan and spec.* This means that the estimator, through the act of submitting a proposal, agrees to comply with the products, applications, and techniques that are called out in the specification manual. No deviation is allowed unless it's expressly granted by the architect and/or owner. Making your own substitutions is *not* acceptable practice.

ABC Contractors

1234 Hammertacker Way, Lienwaiver, WI 53555 • Office (608) 555-1234, Fax (608) 555-1235

Fax Bid Request

To:	Harn Construction	Fax Number:	(555) 541-6786
From:	James Allen	Date:	February 15, 2000

Dear Subcontractor/Supplier:

You are invited to bid on the following project:

Project: ACME addition and remodeling
ACME Manufacturing
Anytown, WI

Architect: Hans Brinker & Associates
Ph. (555) 222-3333 (*Deposit may be required)

Bid due date: Thursday, March 30, 2000 at 12 p.m.

Questions? Call:

Steve Saucerman at (555) 676-2222

Ask for Tracey if inquiring about plan availability

Note: There are four alternates. Please get a copy of the bid form from our office.

Thanks!

File: FAXBID

Figure 6-6
Fax bid request

ABC Contractors

1234 Hammertacker Way, Lienwaiver, WI 53555 • Office (608) 555-1234, Fax (608) 555-1235

Unit Estimate Worksheet

Job name:

Scope of work being bid:

| Plan/ detail no. | Description | Labor | | | | Material | | | | Sub bids/allow ($) | Line total Unit |
		Qty	Unit	Unit $	Subtotal	Qty	Unit	Unit $	Subtotal		
	Totals										

File: UNITEST

Figure 6-7

Unit estimate sheet

ABC Contractors

1234 Hammertacker Way, Lienwaiver, WI 53555 • Office (608) 555-1234, Fax (608) 555-1235

Estimate Summary Sheet

Job name:

Owner/client name:

CSI #	Description	Total labor	Total material	Total equipment	Total sub bids	Line total
	Totals					

File: ESTSUM

Figure 6-8

Estimate summary sheet

However, as you compile the estimate, it's very common to get quotations from subs or suppliers on products or applications that are different from the description in the specification manual. These quotations are called *voluntary alternates*.

It's been my experience that often the company bidding the voluntary alternate may not understand that they've deviated from the specification. It's your responsibility, as the estimator, to insure that the quotes and proposals you receive (and ultimately use) comply with the plans and specs.

Of course, there'll be times when you, or one of your subs or suppliers, know of a substitution that would be of equal or even better value than the one in the specs. There's a formal procedure for submitting these suggestions, involving a written request to the architect. A *request for substitution* is usually submitted at least a week prior to the bid due date, though requirements may differ in your area. It must include detailed information on the proposed substitute product so the architect can determine whether to accept or reject it before the bid date and time.

If the architect, after reviewing the information, feels that the product complies with the intent of the specification, an addenda will be issued to all bidding contractors to inform them that a new product/application has been added to the spec. That levels the playing field — all the contractors have the same opportunity to get a quotation on the new item.

Receiving Subcontractor and Supplier Quotations

If you've been in construction for more than five minutes, you've no doubt discovered the importance of getting a proposal in writing. Verbal promises and commitments often end up confused by "selective memory" as the project proceeds. It's best for all parties to have agreements spelled out on paper. I've included a Bid Quotation form (Figure 6-9) that can help you keep your paperwork organized. You can duplicate it and use it in your business to get started. When filling out the Bid Quotation form, be sure to include the CSI divisions from the spec book for the work you'll be fulfilling

in your bid. It also helps to list the pages from the plans that you used to prepare your bid (A1, A2, A3, etc.). This could help you should the scope of the work you bid ever come into dispute. As a matter of fact, it never hurts to list anything you feel is noteworthy on your proposal — unless specifically instructed not to in the spec manual.

When it comes to the subcontractor/supplier proposal itself, the estimator's most important job is to determine if the bidder's quote covers the complete scope of work. For instance, does the mechanical quotation include gas piping, bath exhaust fans and wall louvers? If you don't see it, or you aren't sure, call and check!

Besides verifying the scope of work, you also need to follow up on these tasks:

▌ Ensure that your sub and/or supplier has seen all addenda, alternates, and unit pricing pertaining to the bid. If they haven't, the bid may not be any good.

▌ If you receive a proposal from someone new, it's reasonable to call and ask about the company and even to ask for references. Good business people realize that you're simply protecting your interests. If they won't comply, or they get defensive, that behavior automatically raises a flag.

▌ Many subs and suppliers like to lump their bid together in one bottom-line number. But when you're doing a bid using CSI line items, you may need a breakdown of the bid. If you do, call and get it. Some companies don't like breaking down their numbers, but if they really want to work with you, they'll cooperate.

Never, ever, "shop" a quotation to another sub or supplier. Shopping, of course, is when you leak numbers to one of your bidder's competitors. Besides the legal and ethical issues this raises, it's just bad business. People talk way too much in this business. Even if it does work once, you'll eventually be found out. When you are, companies will either not bid to you or bid higher to you. Either way, you lose.

ABC Contractors

1234 Hammertacker Way, Lienwaiver, WI 53555 • Office (608) 555-1234, Fax (608) 555-1235

Quotation

To: _____   Date: _____

Project: _____

From: _____   Good for _____ days
from the above date.

Direct phone: _____

We propose to furnish all material, labor and equipment necessary to complete the following scope of work:

_____

_____

_____

_____

_____

For the sum of $ _____

_____

Alternate # ____	❏ Add	❏ Deduct	$ _____	Unit price #1	_____	per	_____
Alternate # ____	❏ Add	❏ Deduct	$ _____	Unit price #2	_____	per	_____
Alternate # ____	❏ Add	❏ Deduct	$ _____	Unit price #3	_____	per	_____
Alternate # ____	❏ Add	❏ Deduct	$ _____	Unit price #4	_____	per	_____

This proposal includes: ❏ Sales tax   ❏ Job site delivery   ❏ Installation   ❏ Winter weather provisions

Exclusions:

_____

_____

I have seen addenda:

_____

Signed _____

Company name _____

File: BIDQUOT

Figure 6-9
Bid quotation form

If your office staff takes phone quotations for you, make sure they know what information you need. Besides getting the base bid number, they should ask for:

1. the company name

2. the phone number

3. a contact person

4. the specific pricing on any addenda, alternates, or unit prices

5. a quote for tax if it's included

6. anything else you consider important to your bid

The phone number is particularly important, especially on last-minute bids. It's often necessary to call a bidder back to check a fact. On a hectic bid day, the last thing you have time for is searching for a phone number! You can put a stack of Bid Quotation forms on everyone's desk, and require that they fill in all the blanks for any phone quotes they take.

Data Compilation and Summary

Once you have all of your in-house take-offs completed, and your subcontractor proposals and supplier quotations accounted for, all you have to do is enter the data and add 'em up. Right? Well . . . kind of.

I believe I forgot to mention that almost all of these sub and supplier quotations will arrive on the last bid day — and many in the last hour before the deadline. Whoever said that people procrastinate? Or, did I mention that many of the numbers on your spreadsheet will change every five minutes? And that the proposals will come in so rapidly and frantically over fax, phone, and in person, that you can hardly remember your own name, let alone remember the difference between the THNN and THWN wire on your electrical quote!

My point is that you need to *stay calm*. Address questions and deal with conflicts as rapidly and as efficiently as possible. A quick answer may not be perfect, but it'll do. When the bid deadline is drawing closer, you need to make decisions and move on.

If you've built your foundation by setting up your spreadsheet completely and accurately, you'll be just fine. Considering the enormous amount of information being assembled and evaluated, you'll be amazed at how well you did after you're done.

As the bid time draws near and your spreadsheet is tallying up the math, your biggest concerns are determining your final markup and filling in the alternates and unit prices, if any are required. When that's done, it's simply a matter of reading the bottom line, taking a deep breath, and letting her fly!

One quick technical note for you spreadsheet users: Make a habit of checking the formulas on your spreadsheet throughout the course of the bid to insure they're calculating correctly. During the last-minute rush of preparing and adjusting the estimate, it's very common to add, delete, and move lines around. Sometimes, when you get going too fast, you can slip up on a "copy" command and fail to carry the formulas that do the actual calculating from one line to another. This is invisible to the user and will cause the line and column summaries to be incorrect. How can you explain what happened to the $24,500 line item that should have been $245,000? Talk about a little stress around the office!

The Bid Form — Presenting the Results

In commercial construction, when it comes to presenting your proposal for bid, the good news is that most of the actual proposal itself is already done for you. There's almost always a bid form supplied in the bid package — often bound in the specification manual.

The bid forms are most often geared towards general contractors, but it's not uncommon to see bid forms for trades as well. This would be the case when there are separate *prime* contracts — where individual proposals are received from general contractors for work in CSI, Divisions 1 through 10, for example, and then the plumbing, HVAC, fire protection, and electrical subcontractors. These take the place of one lump-sum, prime contract (for all divisions) normally submitted by a general contractor. However, in most cases, there's still a general contractor overseeing the bulk of the construction. He generally receives an assignment fee, or something similar, to oversee and coordinate the trades.

Almost all bid forms require the submitter to fill in *all* the blanks on the form. If you have nothing to put in a particular space, don't just leave it blank. Type in *N/A* or *none*. You want to make sure that the people receiving your bid know you've acknowledged each line, or they could reject your bid. The same applies to any alternate lines or unit pricing that may be on the form.

Here's something to watch out for that I've been caught on: Check the addenda received throughout the bid period to see if there's a *new bid form*. This actually happens more often than you might think, and it could mean the difference between having your bid accepted or rejected. While you're at it, you should also check to make sure *you've even received* all of the addenda.

Bid forms range from simple to very complex. I've had anything from one-pagers to proposals that were 40 pages and more. Government, utilities, and large corporations are often the authors of these monstrosities. The additional pages may be filled up with subcontractor listings, minority requirements, bonding and insurance forms, corporate resolutions, prevailing wage compliance, affidavits, and more. When you get one of these, you'll want to fill out as much as possible before the bid date because you won't have time to do it at the end. If you're not sure what's to be filled out, or how, read the instructions. If that fails, call the architect or owner. Don't take chances on doing it incorrectly — it could cost you the job.

The last step in this process is the physical act of submitting the bid itself. Again, you need to follow the requirements. Read the specification. Is the bid supposed to be submitted loose or should it be bound in the specification manual? How many copies are required? Is the envelope addressed correctly according to the spec? Are you sure you're

".... My point is that you need to stay calm. Address questions and deal with conflicts as rapidly and as efficiently as possible. A quick answer may not be perfect, but it'll do."

taking the bid to the right bid-letting location (where the bids are opened)? Is the bid-letting open (they'll be reading the numbers aloud) or closed? When in doubt — check! If it's an open bid-letting, you can attend the reading.

Some of this may seem like fairly petty business, but the last thing you want to do is put a lot of work into an estimate, *only to have it thrown out because of a bid-form technicality*! And that happens quite often. The people who receive these bids aren't kidding around — and neither are your competitors. Why should you be worried about a competitor? Because even if the owner *is* lenient regarding a bid-form error, the competition might not be so accommodating. There are a number of companies who don't take well to losing. They'll go up after the bid letting has finished and review all of the bid forms hoping to find a reason to disqualify the winner — particularly if they're second. *So be careful.*

Conclusion

Construction estimating isn't a discipline that's mastered through reading a book, buying a software program, or listening to someone else describe it. Like most things in life, estimating is a skill acquired through hard work and dedication. Ultimately, the ability to create estimates that are consistent winners will come only with experience and a genuine desire to learn the craft.

True experience ultimately involves some trial and error. I've learned more from my mistakes than from my successes. And quite frankly, *if you're not making mistakes, you're probably not trying hard enough.* Never be afraid to explore new avenues. Nobody likes to make mistakes, but if you use them as tools to help you grow, you'll probably surpass your expectations and reach levels of expertise you didn't think possible. After that, the only problem you'll have is how to actually build all the work you win!

Chapter 7

Owner/Contractor Relationships

Your entry into commercial construction brings with it a whole host of new owner/contractor contractual relationships. Some are similar to what you've already encountered in residential construction, but other forms of agreements may be quite different from those you've seen in the past.

Regardless of their structure, these relationships are the foundation on which your ability to earn a profit rests. If you enter into a lopsided, owner-biased arrangement, you'll suffer financially by the end of the project. On the other hand, if you can negotiate an agreement between yourself and the project owner that's fair and equitable for all parties involved, then you'll certainly end up money ahead.

Types of Commercial Contractor/Owner Relationships

There are several different types of contractor/owner relationships that you can get involved with. Let's look at them.

The Competitive Bid

Because you're new to the commercial construction arena, you'll usually be entering competitive bids — for a while, at least. Competitive bidding is widely used for both private and public projects. The architectural plans and specifications for a project are normally substantially completed by a third-party architectural/engineering (A/E) firm when it's put out for bid. These documents are distributed to the bidding contractors. There are usually at least three companies bidding. Each contractor estimates their cost to complete the project based on the information provided in the documents, then submits a price before the predetermined bid date and time. If the bid is an open letting, the numbers are read in public and generally the lowest qualified bidder receives the work.

Competitive bidding has its good and bad points, which we'll discuss at length later in this chapter. For now, let's just say that the competitive bid process is highly mercenary. The office time expended by the contractor to bid a project can be very costly for a purely speculative venture. Along with the potentially-wasted office time, bidders often fall victim to the intense competition of the bid process. It can pressure them into lowering their profit margins or taking risks that they'd otherwise not even consider — anything to get the bid. There's a danger that getting the bid becomes the sole goal, and they forget that making a profit is what they're in business for. The only positive aspect is from the owner's standpoint. He normally ends up getting a very competitive price for his project.

In a competitive bid, as in other bid arrangements, you may be asked to submit your proposal as either a single-prime or multiple-prime contractor. The single-prime proposal is where one contractor (usually the general contractor) has responsibility for all divisions of the project. This means that he'll collect the bids from all the trades and disciplines required to complete the work as outlined in the plans and specifications, then incorporate them into one large bid. He then presents this lump-sum number as his proposal at bid time.

In a multiple-prime arrangement, more than one contractor will have responsibility for the job. Each one submits a separate bid for his specialty, and is responsible for a specific portion of the project. In a typical multiple-prime situation, a bid proposal is requested for only the general construction portion of the work, with separate bids for the plumbing, mechanical and electrical. Each winning bid will be let as a separate and exclusive prime contract. In most situations, the general contractor will coordinate all the trades once the project is underway, but the owner and architect feel that they have more control of the project with separate contracts. They know how the costs are broken down by trade, and perhaps (although some would argue the case) feel that they're saving some money by not absorbing repetitive markups for those multiple-prime work items.

> *"The main difference between design/build and competitive bid projects is who controls the process."*

Design/Build

Design/build is almost directly opposite to the competitive bid philosophy. Design/build, sometimes referred to as negotiated design/build, is where the contractor and owner strike a single agreement, covering both the design and construction of the project. The two parties negotiate the project, instead of having a group of contractors bid for it. This method relies on many of the positive aspects of the partnering philosophy that's becoming increasingly popular in both public and private commercial construction these days. We'll look at both partnering and design/build construction in much greater detail in Chapter 9.

You'll discover that design/build is a far better construction vehicle than the competitive bid process, but you may have to have a little patience. Opportunities for design/build deals don't usually come along until you've had some experience in commercial construction. That's because the design/build process requires a degree of trust between the owner and the prime contractor that's only achieved over time — usually after working together on other projects. You may find design/build

projects coming your way, even from owners you haven't worked with before, if you've built up an excellent reputation.

The main difference between design/build and competitive bid projects is who controls the process. Though an architect/engineering firm may still be part of the team, the architect no longer has complete control over the conception and design of the project. It becomes a joint effort, with input from the owner, architect, and the contractor. True, the contractor does accept a greater amount of responsibility and liability under a design/build contract, but he can also often count on a slightly better — and more predictable — profit margin.

Fast-Track Design/Build

The fast-track design/build contract is essentially the same as design/build, except the project is started almost immediately upon agreement, often before the plans and specifications have even been fully developed. This is construction on the fly. Fast-track involves more risk due to potential problems that could pop up as the details are fleshed out, but this risk comes with offsetting rewards.

The project is started and completed faster. That can benefit the owner two ways. First, he saves money on interim interest expenses for construction funds. Second, he may benefit by speeding up the transition period from his old facility to the newly-constructed facility. The builder benefits because he can get started on the project quickly, saving time and labor expenses.

Competitive Bid Design/Build

There's a hybrid form of the methods we've just discussed, called competitive bid design/build. This comes about when the bidding contractors are given the end result that the owner wishes to achieve — but very little detail on how to get there. The information given is often very general, though it usually includes physical size parameters, the specific use requirements of the project (such as a medical or institutional facility), and important points or items of particular interest to the owners. The contracting companies bidding the project take this information and work up individual design/build proposals that they feel will appeal to the owner.

Figure 7-1 is a copy of a Statement of Qualification and Letter of Interest that usually precedes the actual cost estimate. It briefly outlines the project and describes the construction company's work philosophy. If the company is invited to build on the project, they will then need to supply a detailed proposal.

The details are up to the bidders. The bidding companies submit a cost for the project based on their own designs. Of course, this means that the bidders aren't bidding on the same thing. The owners look at all the bid packages and then choose a winner based on both price and design. Needless to say, this can be quite an expensive undertaking for the bidding contractors. I don't recommend taking on this kind of bid until you've had considerable experience as a commercial contracting company.

Fee Arrangements

When dealing with any contractual relationship, you'll need to set up a fee schedule. Fees can be structured in a variety of ways, with pluses and minuses for each arrangement. You'll eventually decide on the arrangement that suits your company best, but that'll most likely come through trial and error. Let's look at a few of the more popular fee structures for a commercial construction contract.

Lump Sum or Fixed Fee

The lump sum fee is just as simple as it sounds. It's a set amount of money designated as a fee. It may be $5,000 or $500,000, but whatever it is, it doesn't change during the course of the job. The amount will most often be determined by the size and scope of the project.

Percentage

A percentage fee is based on a percentage of the total overall construction cost. It's a very common fee method, and the one that you'll probably encounter most often. The amount of percentage used can vary with the size of the job, market pressure, your own financial needs, or other factors.

Cost-Plus Percentage/Cost-Plus Fixed Fee

A cost-plus fee often involves an "open-book" arrangement where the general contractor shows the owner the actual costs of the project as they're received. The general contractor then gets paid the project cost (to cover the actual expenses), plus a fee based either on a percentage of the total or an agreed-upon fixed amount above that total cost. A budget estimate is prepared at the beginning of the project, and the ultimate goal is to have the project come in on schedule and on budget. If the real costs in a cost-plus arrangement run over budget, the additional cost usually doesn't come out of the contractor's fee. That fee is pre-set (although other arrangements can be specified in the contract). In this sense, a cost-plus arrangement takes some of the pressure off the contractor. He knows he'll get paid, regardless of whether the project comes in on budget or not. An example of a cost-plus percentage contract is shown in Figure 7-2. You can use a similar contract for a cost-plus fixed fee contract. You just need to change the wording slightly under item number 3 to read: The owner agrees to pay the contractor for the actual cost of the work plus a fixed fee of _____.

Cost-Plus Guaranteed Maximum

A cost-plus guaranteed maximum contract is nearly the same as the other cost-plus arrangements, except here the parties agree to put a cap on the amount spent. The positive side is that the owner knows the job won't exceed the funds he's set aside. The down side is that the owner and the construction manager may need to do some cost cutting or provide some other remedy to keep costs within the agreed-upon maximum.

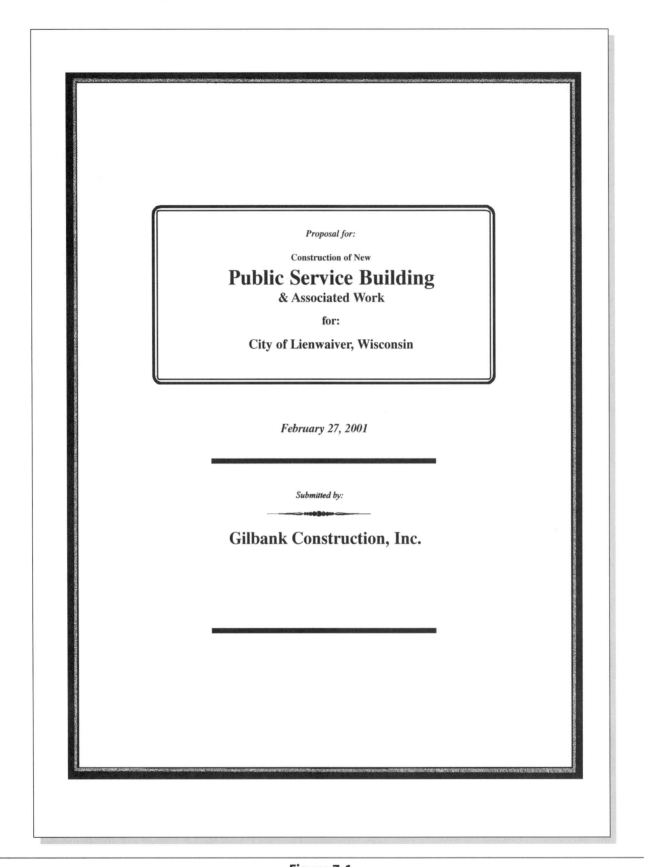

Figure 7-1
Statement of Qualification and Letter of Interest

GILBANK CONSTRUCTION, INC.
GENERAL CONTRACTORS
Commercial • Industrial • Residential

Established 1965

February 27, 2001

City of Lienwaiver, WI
Lienwaiver, WI 61080

Attn: John Smith — Mayor
Re: Letter of Interest

Dear Mayor Smith:

Thank you very much for the opportunity to submit a Statement of Qualification and Letter of Interest on design/build services for your new public service building. Please accept this letter as affirmation of Gilbank Construction's sincerest interest in the proposed project.

We've enclosed a contractor qualification package which includes an overview of the project, general and financial information on our company, and also a sampling of some of our past work — any of which we invite you to view at your leisure. Regarding the build/lease note in your letter, we currently are entered into build/lease arrangements with other public entities and would be very open to discussing the possiblities of this type of arrangement with you.

Gilbank Construction will perform all of the duties normally associated with quality design/build commercial construction including:

- Assistance with planning and development

- Complete architectural & engineering services

- Preparation of budgets and preliminary cost projections

- Solicitation of sub-contractor/supplier proposals and development of final cost estimates

- Prompt, supervised, and safety-conscious construction throughout the project

- Accounting and administration

- Thorough closeout, warranty, & O&M procedure

- Courteous, professional, & responsible service during and after construction

Working in cooperation with the City of Lienwaiver, our goal and promise is to deliver a cost effective, quality, and on-time construction of your new public service building. Please call if you wish anything further or have questions regarding this proposal. We look forward to hearing from you.

Sincerely,
GILBANK CONSTRUCTION, INC.

Steve Saucerman

Enclosures

Figure 7-1
Statement of Qualification and Letter of Interest (continued)

City of Lienwaiver Proposed Public Service Building (cont.)

Project Scope

The project scope includes the design and construction of a new Public Service Building for the City of Lienwaiver, WI. The proposed location for the project would be lot no.32 of the Lienwaiver Industrial Park (fig.1) and the construction would include all necessary public utility tie-in and connections. All work performed will comply with all governing federal, state, & local codes and the Americans with Disabilities Act (ADA).

The proposed building is an approximately 14 M sf single story , slab- on-grade construction including insulated CMU exterior walls, brick veneer on front elevation & no-maintenance siding on side & rear elevations. All mechanical & electrical as necessary including a natural gas forced-air heating system in main building, unit heaters & radiant panels in garage bays, and ADA accessible plumbing installations.

Lot 32 — Lienwaiver Industrial Park

Scope of Services

Gilbank Construction offers complete and professional construction services for your project — from conceptual estimating to the completion of the punchlist, including:

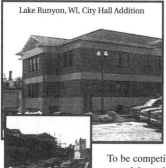
Lake Runyon, WI, City Hall Addition

Bidding & Costing

Gilbank Construction provides all services associated with competitively bidding a construction project. Services include, but not limited to:

* Sending out bid invitations to sub-contractors & suppliers that we know and trust

* Advertising in construction publications .

* Creating value-engineering suggestions for the owner .

* Distributing bid & construction documents to the bidders .

* Receiving / compiling bid proposals and verifying scope of work

To be competitive, we will always attempt to solicit at least three bids for each major trade or material supplier. Proposals are then compiled on computer in CSI (Const. Spec. Institute) format.

Project Management

Gilbank provides comprehensive supervision, scheduling, quality-control, and site-specific safety programs for your project. We maintain clean and professionally managed jobsites. Job progress meetings are held during the construction to facilitate communication among all parties. Thorough logs and documentation are kept throughout the course of the project.

Administration

Gilbank Construction offers complete computerized accounting, record-keeping, & scheduling services for your project. All job costing, lien waiver collection, purchase orders, and sub-contractor agreements are created and administrated through our office in Clinton, WI.

Carro, WI, Village Hall Addition

Figure 7-1
Statement of Qualification and Letter of Interest (continued)

Partnering

Gilbank Construction advocates and encourages the *Partnering* concept, a management process which helps ensure successful project development and execution. The concept is important because every contract and agreement between the parties includes an implied covenant of good faith. While a written contract would still institute legal responsibilities, *partnering* attempts to establish a productive working relationship between Gilbank and the City of Lienwaiver through a mutually developed, formal strategy of commitment and communication, creating an environment where trust and teamwork minimize disputes, foster cooperation and facilitate completion of a successful project for all.

The following is a typical scope of work outline for this type building:

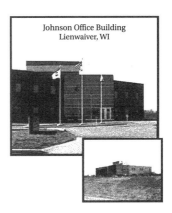

Johnson Office Building
Lienwaiver, WI

Project Scope of Work (By **CSI** Division)

CSI	Description
2	Sitework, Excavation, Underground Utilities, Site Concrete & Asphalt Paving
3	Concrete Formwork, Cast-in-place Concrete, Concrete Reinforcement
4	Masonry Block and Brick Veneer Work
5	Structural Steel & Miscellaneous Fabricated Metal Products
6	Rough & Finish Carpentry
	Architectural Woodwork & Cabinetry (Those Items Not Provided Directly by Owner)
7	Foundation Dampproofing
	Thermal & sound insulation
	Roofing, Sheet Metal, & Roof Accessories
	Caulking & Sealants
8	Hollow Metal Doors, Frames, & Hardware
	Architectural Wood Doors
	Overhead Doors, Accessories, & Operators
	Aluminum Framing, Glass and Glazing
9	Gypsum Wallboard Systems
	Acoustical Ceilings & Grid Systems
	Quarry Tile, Resilient Flooring, & Carpeting
	Painting, Staining, & Special Finishes (If required)
10	Toilet Partitions, Accessories, Miscellaneous Specialties
15	Plumbing & Piping
	Fire Protection & Sprinkler Systems
	Heating, Ventilating, and Air Conditioning (HVAC)
16	Electrical, Communication, and Lighting

Figure 7-1
Statement of Qualification and Letter of Interest (continued)

ABC Contractors

1234 Hammertacker Way, Lienwaiver, WI 53555 • Office (608) 555-1234, Fax (608) 555-1235

Cost-Plus Percentage Contract

ABC Contractors
1234 Hammertacker Way
Lienwaiver, WI 55555

and

Mortimer's Grocery
555 Main Street
Anytown, WI 55555

An agreement is made this _____3rd_____ day of _____July_____, 20_01_ between the parties above whose terms and conditions include:

1. The contractor and owner agree to enter into a formal construction partnering arrangement founded on mutual trust and respect for the other party. The contractor hereby pledges service and skill to the best of his abilities in order to facilitate a successful project and further the interest of the owner. The contractor further pledges to supply the owner with efficient and professional job administration, accurate accounting, skilled craftsmen, quality materials, and a workmanlike manner.

2. Work shall begin _____August 1, 2001_____. Work shall be substantially complete by _____November 30, 2001_____. There ~~are~~/are not (strike one) liquidated damages or other monetary penalty in the amount of $ ___N/A___ per calendar day that the project extends past the substantial completion date.

3. The owner agrees to pay the contractor for the actual cost of work plus a negotiated percentage of _9_ % based on the actual cost. The "cost of work" includes all labor, materials, equipment, equipment rental, and service related to the completion of the owner's project. This includes all changes and modifications to the contract which occur after the project begins that are agreed to by both parties.

4. The scope of work for the project is attached to this agreement and shall become a part of the agreement. Any future changes in scope shall become attached to this agreement at that time. A "good faith" estimate has been submitted by the contractor in the amount of $ ___360,000___ to be used as a monetary guideline, but that estimate shall not dictate the final cost of work. All overages shall be paid for by the owner and all savings on the project (under the estimate) shall be returned to the owner.

5. The contractor is responsible for obtaining all normal indirect and/or general condition work items such as insurances, permits, bonding, governmental fees, and so on.

6. The contractor agrees to perform the following work "in-house" (with his own crews): _____ _____*excavation, concrete & carpentry*_____, the remaining work shall be subcontracted to reputable, skilled companies.

Figure 7-2
Cost-plus percentage contract

7. The contractor agrees to share the accounting of the project in an "open-book" fashion with the owner. This entitles the owner to copies of all invoices and labor worksheets. In addition, the contractor will furnish to the owner ___*one*___ time(s) per month a progress billing showing individual work items and their percentage complete for that billing period.

8. The contractor will carry and maintain necessary workers' compensation and liability insurances as to protect the owner from loss and/or liability during the course of work. In addition, both parties agree to hold the other harmless and to indemnify each in the event of a dispute or claim raised during the course of construction, unless that dispute or claim is the direct result of a deliberately negligent act upon the purveyor. This hold harmless agreement shall apply to all companies, employees, and their directors.

Therefore, this agreement is executed this day and year as described above:

John R Smith 7/3/01   _A.L. Johnson_ 1/3/01
Owner Signature   Date   Contractor Signature   Date

WI Lic. # ABCD000W
Contractor license number (if applicable)

Figure 7-2
Cost-plus percentage contract (continued)

Cost-Plus with Savings

Again similar to the first cost-plus arrangement we discussed, cost-plus with savings offers the opportunity for the prime contractor to split with the owner any savings that are realized when the job is closed out. That's in addition to his fee. The percentage of this split (50/50, 60/40, etc.) is negotiated prior to signing the agreement. This can be a very attractive fee arrangement because it offers an incentive to both parties to remain lean and cost-effective during construction.

Other Fee Arrangements

The line between contract types and fee arrangements has been blurred over time. It's not at all unusual to see different combinations of the situations we've discussed. In the end, the agreement (contract) formed between you and an owner is limited only by the willingness of both parties to accept the terms as stated.

The Problem with Competitive Bids

My view of the competitive bid arrangement is probably a little different from what you've heard from other contractors or read in other books on construction contracting. I'm sure, if you've been in this business any time at all, you've formed your own opinion on this subject. You know that it's the most frequently-used agreement between an owner and contractor. In spite of its apparent popularity, I feel that the arrangement is flawed. Unfortunately, my opinion may just sound like whining — the unhappy result of having come in second in too many bids. Despite this, I'll try to justify my statement.

The fundamental problem with the competitive bid is the process itself. I'm not the first one to notice that this process has "issues." It's difficult to pick up any construction industry publication without finding some discussion on the competitive bid and its shortcomings. The articles almost always end up praising the benefits of negotiated and design/build work over competitive bid projects. All the authors have their own special prejudices, and

convictions, but there's one common thread that binds all of the discussions together: the element of trust.

There's a general lack of trust between the players in a typical construction deal — the project owner, the contractor, and the architect. We say we trust each other, but we don't. Why? Well, there's a somewhat tawdry history of builders, owners, and designers doing end-runs around each other in the hopes of pumping up their own profits. It's happened enough for each side to have legitimate reasons for being cautious when it comes to the motives and intents of the other parties in a construction deal.

Although admittedly biased, I'll attempt to show you how your trust can fall prey to the elements of the competitive bid process. If you're just entering the commercial field, maybe a word from a battle-scarred student at Hard Knocks University will save you a few bruises. I'd like to say I'm a graduate, but I'm afraid I still keep getting educated! Let's walk through the bidding process from the beginning.

The Bid Process

Using identical bid packages, each competing company will analyze, estimate, and agonize over the documents in an attempt to arrive at a true and competitive price for the project. Before the deadline, the companies deliver their completed bid forms to the owner's representative.

Once received, the bids may or may not be read aloud. If they're read aloud in an open letting, each bidder knows, at least from a pricing standpoint, where he stands. At a closed letting, the bids are opened by the owners or their representative (often the architect) in private session at an unannounced time and location. The results may or may not be made public to the bidders. It can be days, weeks, or even months before the bidders are notified of the results. Sometimes they're never notified.

These occasional long waits are frustrating to the builder. You have to schedule manpower based on your known work load. This simple failure to communicate with the bidders, either yes or no, is the first step in the erosion of trust between the contractor, the owner and architect. It shows a basic lack of respect for the effort that the contractors have put into their bid proposals.

But even an open letting has its perils. Having the low bid at the letting, although generally a positive sign, doesn't necessarily assure a builder of getting the job. There's almost always some type of wording in the bid documents that gives discretionary power to the owner to choose the contractor that best suits the overall needs of the project. The result is that the owner and architect can simply choose whichever contractor they please, for whatever reasons they feel are appropriate. All the bid process does is force the contractors into competition and drive the contract price down.

You may say, "Ah, c'mon, no system is perfect. The owner has his money on the line — he's got to have some control over his fate! The competitive bid provides for healthy American competition. Right?"

Well, competition yes; but it's not exactly healthy. There are also some other problems with this system.

The Owner/Builder Relationship

From a relationship standpoint, the competitive bid is devastating. Why? Because the moment the owner selects competitive bidding as the vehicle for establishing his future contractual relations with a builder, these results are almost certain:

▪ The relationship between the owner, the architect and the builder will become adversarial. The major parties involved in the contract are now in a position of financial competition, each protecting his own stake in the building project. Each one wants his share of a limited amount of money, and whoever can hold onto the most wins! Budget mishaps do occur in design/build situations, but in a congenial atmosphere, the parties work together and come up with solutions. In competitive bid situations, this isn't the case. Lines are drawn in the sand, sabers can rattle, and it can get very ugly!

▪ There's no bond or trust — no sense of partnership in the venture. The construction project is just a business transaction, and if one party suffers in relation to another's good fortune, it's justified in the minds of all the players as "protecting one's own interests." And this callous detachment works both ways — contractors can be just as cold and unsympathetic to the owner's problems as the owners are to the contractors'. However, knowing that's the case doesn't make working under these circumstances any easier.

▪ If the project goes smoothly, the relationship may remain civil, even friendly, and they may do business again. But the quest for profit and the advancement of one's own agenda are always first and foremost in this arrangement.

Bid Documents

Most people assume that in a competitive bid situation the contractors receive and bid from complete and professional construction documents. I'd personally estimate that perhaps 10 percent of the plans from which I prepare my bids are complete enough to generate a true "apples-to-apples" competition among the bidders. About 15 percent are totally inadequate, and the remaining 75 percent are somewhere in between. This means that more often than we'd care to admit, we estimate the cost for the project without a thorough and precise understanding of the construction documents. This is hardly what the process was intended to produce!

I've found myself involved in competitive bids where, due to the incoherence or incomplete state of the bid documents, I had to arrive at my bid number through my own intuition. In other words, I have to use my experience as a builder to determine the "intended" scope of work. I imagine that the other estimators are doing the same thing. And just what are the odds that we're all interpreting the bid documents the same way? Even though we're all working with the same documents, we may be seeing something entirely different. That situation doesn't produce a truly competitive bid. I think that the lack of quality and completeness of the architectural plans and specifications is one of the biggest obstacles to the success of the competitive bid process.

Why are we seeing this problem so often with architectural work today? The architectural schools are putting out graduates just as skilled as they ever did. Perhaps it's market driven — the result of increased competition between architectural firms. They have to spend less time on a project in order to keep the cost competitive. I used to place architects on a higher level, well above someone like me who works with his hands and gets mud on his boots. But now, being in a position where I have to actually figure out the job from their plans, I realize that they're no better than the rest of us. What a disappointment!

I've also come to realize that there's a huge difference between the best and the worst in the architectural field. Again, is this from increased pressure on the job? Like contractors, they have to outbid other firms to get work. Perhaps they just can't give each job the amount of time it needs. Whatever the reason, builders like you and me have to deal with it, and it isn't easy.

So what do you do when you're confronted with incomplete documents? Well, somewhere in the specification manual it says that we should notify the architect of any mistakes, omissions, or discrepancies that we find. But sometimes there are so many errors and omissions on the plans and specs that we just try to work around them. Some of us do, however, try to rectify errors with the architect during the bid process. But even when you do that, most of the time you still won't get it corrected. The architects are probably on another high-priority job and aren't about to stop and miss their deadline in order to put in some detail on a job they already finished. Or else they're embarrassed, but don't want to admit it. Either way, don't hold up your bid waiting for them to call you back. There's a good chance you'll be waiting till you retire. Just make your best guess and move on with it.

Honesty in the Trades

There are builders who believe that omissions in the bid documents are actually premeditated. They think it's a trick by the owner and architect to get an inexperienced bidder to bid substantially lower than the rest of the bidding contractors, so they can legally and contractually force him to perform the work at a loss.

I don't buy into this theory, but many builders swear this happens. There's even a joke about it. They call it the "who-missed-what" method of construction award, or "Congratulations?? You're low!!"

I think that the blame belongs to all the parties, including the builder. We've allowed ourselves to be slowly and methodically drawn into an award process that forces us to employ less than desirable (and sometimes less than ethical) strategies to achieve "success."

The Competitive Difference

Despite what many people believe, contractors have very little competitive advantage over one another when it comes to cost. With minor exceptions, we all purchase the same building materials, labor, subcontractors, and equipment for about the same outlay. Most of the time, the difference between winning and losing is more a product of minor differences in overhead, profit margins, work load, and just plain aggressiveness.

There are no tricks, no warehouses full of deep-discounted materials, and no mystical technical knowledge that distinguishes one honest builder from another. Indeed, if you're searching for a difference, your time would be better spent examining attributes such as integrity, experience, and reliability.

I've been to a lot of bid lettings, and sometimes there's a tremendous spread between the lowest and the highest proposals. This should immediately throw up a red flag for the owner and signal potential problems with the bid documents. But a lot of them just don't see it.

Unless he conducts a thorough prescreening of the bidding competitors, an owner can find himself with huge variations in quality, experience, and competence among those participating. In a situation like this, the owner will certainly get what he pays for!

It's quite expensive, in terms of office overhead and employee time, for builders to prepare estimates for competitive bids. That time has to be accounted for someplace. Most contractors work it into their overhead and spread the cost over all their jobs, but that just increases the cost of doing business in general. I've read articles proposing that owners offer a small payment of some kind to all those bidders invited to bid their projects to help cover some of the overhead expense. Although I think this is an excellent idea, I've never heard of an owner actually doing it.

Many established building companies simply don't do competitive bidding anymore. And why would they? They have repeat clients and an ongoing relationship with the business sector in their community. Since these experienced contractors aren't included in the "bidding pool," the owner who sticks to the competitive bid process loses access to a lot of quality builders.

There are many benefits of dealing with seasoned professionals. One of them is value engineering, which allows for the review and replacement or elimination of items that are excessively expensive or unnecessary before they become part of the actual building package. There are ranges of quality in a large proportion of the line items on a construction estimate, and many different products and applications that will reduce the overall project cost. Experienced contractors easily pick up on these items. Why don't the owner and architect practice value engineering? Because they simply don't have the builder's expertise and knowledge when it comes to the cost of material, labor, subcontractors, and equipment. Builders work with these costs all the time — day in and day out. Most architects and owners don't.

But the best reason for working with an experienced professional builder in a noncompetitive atmosphere is that the owner, architect, and builder become a team. Once the strife inherent in the competitive bid is eliminated, the players work together to achieve a successful and profitable construction project for everyone involved.

If Not Competitive Bid — What Then?

For the builder, the most popular alternative to the competitive bid is a negotiated design/build deal. (We'll look at this method in detail in the next chapter.) In this situation, the owner works with a builder from the beginning to create the project. If the builder doesn't have in-house design services (and most don't), an outside designer may be brought in as part of the team.

The important difference in this type of project is that decisions are made in cooperation with one another, and each participant has a say. The relationship is built on trust. As a gesture of good faith, the builder often makes his accounting books available to the owner or the owner's accountant for review. In this way the owner is assured that he's getting the best possible value for his money. The builder is content because he knows he's negotiated a fair fee for his work, and the architect should be content as well. Unfortunately, you're probably not going to find many architects who'll recommend this type of arrangement. Many view design/build deals as an assault on their turf.

Acquiring design/build projects really is about getting the owner to trust you with his deal. You have to sell the concept. You have to explain the benefits and point out the disadvantages of the competitive bid process. Otherwise, most owners will just stick with what they've always done. Getting an owner to trust you doesn't happen overnight. Usually this kind of trust is build up over a series of smaller work projects and past associations and recommendations. As your company and your reputation grow, more opportunities to sell the design/build idea will come your way.

I believe that this type of working relationship — built on respect, trust, and honest negotiation — is the only avenue that affords real value to all parties. And isn't value what it's all about?

Chapter 8

Design/Build and Partnering

D esign/build is a strong alternative to the competitive bid. When you combine it with aspects of the partnering concept, you not only end up with a good alternative, you end up with a better building project.

In his autobiography, Mark Twain said, "There are three kinds of lies: *lies, damned lies, . . . and statistics."* — so it's with some hesitation that I begin this discussion with a statistic. However, the numbers revolving around the design/build phenomena are certainly worth noting. According to the *Design Build Institute of America* and a feature article in a recent *US News & World Report*, the use of design/build has enjoyed a steady and rapid climb among construction projects performed over the last few years. And it continues to grow in popularity. In fact, both sources estimate that design/build now makes up about 30 percent of commercial construction contracts, and they estimate that it'll almost surely rise to about a 50 percent market share by the year 2005.

What Is Design/Build?

Design/build is a contractual relationship in which an individual general contractor works directly with a project owner in all phases of a building project. The team, which often includes an architect

or designer during the planning and design phases, works together from the very beginning of the project to achieve a common end. They define a budget for the project and once it's established, they're obliged to do whatever it takes to keep the project costs within those parameters.

The early design phase includes the same elements that you'll find in any contract arrangement, except that the contractor is part of the process from the beginning, and helps to establish the budget for the project. As with competitive bids, he still has to evaluate the job, make estimates, and get bids from subs and suppliers for their portion of the project. However, because they're using his budget figures, he's comfortable with the costs.

Having the contractor in on the planning and development stages helps keep the cost estimates realistic. A budget established by an architect working alone is often grossly underestimated. As a matter of fact, I've been involved in competitive bid situations where the actual bid numbers are anywhere from 10 to 100 percent over the architect's budget.

As a team, the owner, builder and architect work together to create and refine the plans and specifications, and define the cost estimates. Special engineering considerations (mechanical, electrical, structural and civil) are targeted and addressed, and zoning and code adaptations for the project are hammered out. As the details come together, the projected construction schedule also begins to take shape.

The owner and contractor can establish a comfortable working rapport knowing that the cost element is out in the open and under control. Open communication fosters the give-and-take atmosphere that's essential for the design/build project to be truly successful.

The Pros and Cons of Design/Build

You may ask yourself, "If it works so well, why doesn't everyone use design/build?" As in most things, there are *pros* and *cons* associated with the design/build process. Figure 8-1 shows a summary of some of the design/build pros and cons. In order to make an informed decision on what's right for you, you need to sit down and weigh the good aspects of design/build against the less favorable aspects of this type of contractual arrangement.

Design/Build Pros

In general, the favorable aspects of design/build are good for everyone as long as each member of the team treats the others fairly. Here are some of the more important benefits.

▌ Design/build fosters a team atmosphere, eliminating the "*me vs. them*" attitude. The owner, architect, and builder make a commitment early on to work together, as partners in the venture, to find solutions to problems. Design/build relationships are conducive to open debate and brain-storming in order to come up with solutions to difficult situations. With greater openness and communication, conflict and stress are reduced.

▌ Design/build reduces blame-passing. The builder can't blame the architect for a poor design when he was a part of the design process. No one can complain about the budget if they've worked it out together. And best of all, the owner isn't caught in the middle of disputes between the architect and builder.

Design/Build Pros and Cons	
D/B Pros	**D/B Cons**
Process fosters team atmosphere	Owner may leave himself open to advantage
Value engineering, open debate and brainstorming may save owner money	Owner loses benefit of competitive bid to drive general contractor costs down
Promotes early detection and remedy of design errors	Architect loses some of his control over project
Reduces blame-passing	Contractor accepts more responsibility and liability
Greater project and document flexibility	Construction documents need to be explicit regarding owner/architect/builder relationship
Owner avoids the double markup	General contractor could potentially receive a higher fee
Change orders are minimized	General contractor may have to be licensed as a designer
Overall attitude is improved for all members of the team, which promotes a pleasant working atmosphere	Developing trust can take time, so design/build may not be available to less-established contractors
Allows for a more civil parting of owner and contractor should disputes arise	Law may *mandate* competitive bid for some projects
Owner has a single source of responsibility — the general contractor	Contractor must take responsibility for any design errors since architect works under him
Low-balling is eliminated	Overall cost to the owner may be higher

Figure 8-1
Summary of design/build pros and cons

▌ Any errors or conflicts in architectural and engineering designs are discovered early and corrected, eliminating costly delays and requiring only minimal alterations in the documents. Specifications are also tailored more closely to the project and include fewer "boilerplate" specifications (those specifications that are simply repeated from one spec book to another, and over time, become inappropriate or just plain obsolete).

▌ Job change orders are minimized. Again, from a contractor's standpoint, it's less likely that you'll need to make changes if you were part of the design process. If you do need to make a change, design/build provides a far more agreeable atmosphere for negotiating that change with the owner. (But it's also harder to ask for a change that requires an additional charge if the need arises.)

▌ There's greater contract flexibility with design/build. You can negotiate anything you feel is necessary into the contract, and use a far more non-partisan document than the *American Institute of Architects* documents conventionally used in competitive bids. I've found the AIA contracts and documents tend to favor or protect the architect rather than the contractor. Of course, I'm looking at it from a contractor's viewpoint. But in my experience with bid situations, the process is in full swing by the time the contractor gets the contract. That leaves the contractor with the option of rejecting any lopsided documents and possibly losing the job, or accepting the package set before him. I don't know how we contractors let ourselves get into a situation where our contracts are written by architects, not contractors!

▌ Unfortunately, sometimes things just don't work out. Design/build contracts can be structured to give all the players certain "outs" should the project begin to go sour. You can include contract wording that allows for a civil parting of the ways, often with instructions as to how payments should be made up to that point. Though this rarely happens, it's nice to know that there are options in place if it becomes necessary.

▌ The owner may be able to avoid the *double markup*. It's common in competitive bid situations for the general contractor to automatically mark up the entire bid before submitting it, even though a sub or supplier proposal has already been marked up by that company. With design/build, the sub and supplier quotes are generally shared openly with the owner, and these may or may not be marked up again by the contractor, depending on his arrangement with the owner.

▌ The contractor has a good attitude toward the project and is willing to share cost-saving ideas, lowering the overall cost of the project. With design/build, value engineering is a continual part of the process. They'll use commodity or standardized products instead of top-of-the-line and single-source products. A side benefit of value engineering is the higher level of craftsmanship you'll get when using standardized products. When your crews and subcontractors are installing familiar products, they make fewer mistakes. Fewer mistakes mean fewer corrections, fewer schedule interruptions, decreased warranty problems due to faulty installations, and an overall savings in labor costs.

▌ With design/build, the owner enjoys a single source of warranty responsibility. Since the architect works under the contractor, it's no longer a matter of design error or construction error. The owner simply picks up the phone, calls the contractor, and says, "*fix it!*"

▌ The overall project schedule (which includes planning, design, contract negotiation, the actual construction of the project, and project closeout) is shorter with design/build. This is primarily due to the fact that the contractor gets a running start on the design phase.

This can be even quicker when using *fast-track* design/build, which allows for the site work and foundation portions of the project to be started even though the plans and specifications aren't fully developed. Fast tracking also enables the submittal and shop-drawing process to start earlier than in a competitive bid situation (where contracts must first be signed). That means the team can release special-order and long lead-time items before ground is actually broken. These items can hold up a construction schedule and cause delays in completion.

▌ Design/build helps to eliminate the *low-ball* bid. Low-balling is purposely and artificially bidding low in order to win a construction job. As we mentioned in the last chapter, a contractor may do this with the idea that he'll be able to make up the difference in costs with future change orders.

▌ The last benefit is that the building process is more fun! I know we're not in this business to entertain ourselves, but when given the option, wouldn't you rather be working with someone you like? Isn't working in a congenial atmosphere better than working with someone you don't like, who you know is going to try to take advantage of you the moment you let your guard down?

Design/Build Cons

Most of the negatives associated with design/build affect the owner and architect far more than the builder. It would be fair to say that design/build arrangements favor the contractor.

▌ With design/build, the owner loses the opportunity to collect and compare three to five (or more) general construction bids for his project. Eliminating bids also eliminates the market competitiveness that's part of the competitive bid process, and may result in a slightly higher price on the actual construction for any given project.

▌ In the atmosphere of mutual trust generated by the design/build arrangement, the owner potentially leaves himself open to abuse by a less-than-honest contractor. Though I hate to admit it, there are contractors who would take advantage of this situation by puffing up prices, using substandard workmanship, or just generally skirting around proper construction process. This is a particularly hard situation to recognize and correct if the owner lacks construction knowledge. And it hurts us all.

▌ Since the architect is employed by the contractor rather than the owner, his contact with the owner and his say in the project tend to fade into the background. This gives the general contractor greater latitude in running the show, creating an additional risk to the owner. Even an honest contractor can make mistakes. When the architect works directly for the owner, he doggedly monitors the owner's interest throughout the project. Without this check on the contractor, the owner may conceivably end up with less of a product than he'd originally intended — through deficiencies in detail, specification, installation procedures, and quality.

▌ What happens to the architect if the owner fires the general contractor? Is the architect, who's employed by the contractor, fired too? What happens to the construction documents? These questions need to be addressed in the contract when it's written, not left to be worked out if things should fall apart.

▌ When the architect is employed by the general contractor, the owner can lose strategic legal and insurance advice. There are a number of preprinted contracts and supplements published by different professional construction organizations available for owners and contractors to use for design/build projects. But there's still a need for each contractor to ensure that the legal and insurance requirements in his state are interpreted correctly in the documents.

When an owner employs an architect, the architect provides insurance for architectural and engineering errors. The general contractor probably doesn't carry this type of insurance, and the architect may no longer want to provide it if he doesn't retain sole control over the construction documents. There's also the question of whether bonding protection should still be applied to the project. These items become clouded when the architect is an agent of the contractor rather than the exclusive agent of the owner.

Depending on the type of project and where you live, there may be laws that require an open competitive bid, or that require an independent designer be employed on every job. In some areas, a contractor may need a special license to perform design/build work if the architectural/engineering work is done in-house. Courts in certain jurisdictions have found design/build contracts to be illegal and/or unenforceable, particularly if the contractor isn't a licensed design professional. Before you contract for this type of work, check on the laws in your area and be sure you understand your responsibilities and requirements.

In order for a design/build project to be truly successful, an owner must have confidence and trust in the contracting company he employs. Although it takes time for this type of relationship to develop, it's essential if the design/build concept is to work.

Unfortunately, it takes newcomers to commercial construction a while to develop a reputation that'll lead to these quality relationships. So this type of work may be presently unobtainable for you — and that can be a major drawback to design/build. But have patience; these jobs will come in time.

Why Hasn't Design/Build Replaced the Competitive Bid?

You may ask, "If design/build is so great, why are the majority of construction awards still the result of competitive bids?" The answer to that question probably involves several factors, from the desire of architects to remain in control of the construction process, to that all-too-human condition of old habits dying hard. I'm sure that you've heard this said a million times: "But that's the way we've always done it!"

There's no doubt that design/build is a concept you'll have to sell to an owner. Design/build customers don't come along everyday, and all projects don't lend themselves readily to this type of arrangement. But when the right customer does come along,

you'll know it's time to introduce this concept. I'll also wager that the client comes to you because he *trusts* you . . . and you *respect* him. And that's the way it should be.

Partnering

You could call *partnering* an extension of design/build. The partnering concept carries with it all of the attributes of design/build, yet it also brings an innovative approach to the essential element of construction administration. But what is partnering? Simply stated, partnering means "*joining or associating with another.*" A more precise definition for our use in construction would be "*a construction management philosophy that attempts to improve project development, mobilization, and management through structured teamwork among all parties.*"

The idea of partnering came out of studies conducted in the construction industry in the early 1980s. There appeared to be a growing lack of faith between parties involved in construction projects — and a definite need to correct the situation. Construction had become a tension-filled industry progressively undermined by lawsuits. The concept of partnering was seen as a way to get back to a less-confrontational business attitude, and create a working atmosphere that promotes good will and trust in a win-win situation.

The partnering concept calls for all the parties involved in a building project (the owner, general contractor, subcontractors, suppliers, architect, etc.) to sign a *formal written agreement* stating that they'll work in an atmosphere of trust, good faith, open communication and commitment to the project. The goal of this declaration is the elimination of the disputes and misunderstandings that often plague construction projects, leading to costly claims, litigation, and/or bad feelings among the parties.

In addition to this written covenant, the details of the project, including construction costs, value engineering, contractor fees, scheduling and dispute resolution, are discussed openly. These details, once mutually agreed upon, become part of the overall agreement. The who, what, why and where of the discussions are entirely up to the group involved in the negotiation process. The important factor is that everyone comes to a mutual understanding of what's expected of each other.

Why Partner?

The theory behind partnering is certainly a good one. It's an attempt to counteract the negative past experiences of owners, architects, contractors, suppliers and customers. We all know that people (any one of us) can be fickle at times — and given the number of personalities and temperaments involved, a construction project can head south in a limitless number of ways. Partnering isn't going to eliminate all misunderstandings, but even eliminating half of them is better than the alternative.

"The name of the game is "no surprises," and to reach that level, it's essential that you have clear communication."

If you need convincing, sit down some time with an experienced commercial general contractor, an owner's construction representative, or architect/designer and ask them about some of their jobs that didn't go *quite like they had planned.* Then get comfortable — because you're going to be there for awhile.

You'll hear tales of corrupt contractors, bumbling architects, and lunatic owners (depending, of course, on who's telling the story). With a few minor variations, the stories will be the same, but the storyteller will *always* be the good guy. There will be wrongful change orders, poor documentation, broken promises, and claims that "the other guys just didn't understand our side." If you listen closely, you'll soon figure out that the real reason that jobs went wrong was communication — or rather, lack of communication. No one was really good or bad, they were just a bunch of guys who weren't communicating. *Partnering* attempts to correct, or at least improve, these types of problems.

When to Use Partnering

On paper, any job can be partnered. When dealing with real people, however, you need to employ a little common sense. Partnering takes a lot of face time

with the owner — more than many contractors are used to. The name of the game is "no surprises," and to reach that level, it's essential that you have clear communication. This involves preliminary meetings, bid meetings, post meetings, emergency meetings, meetings to schedule other meetings, and so on. You get the point. It can take up a lot of time and extra effort by you and your staff to make it work.

For each situation, you need to analyze the cost of your time and weigh it against the potential profit. Office time costs money, too; the same kind of money that you pay a carpenter or an electrician. If the total job cost is only $25,000, and your negotiated percentage or fixed fee is $5,000, you're clearly not going to be able to devote a lot of man-hours to developing the partnering concept. On the other hand, a $3,000,000 contract (with a potential $250,000 fee) will make the office overhead costs a little easier to justify. You might also consider that the office time spent up front in partnering sessions could help prevent a lot of office time spent battling over budget problems or other job concerns that haunt many projects.

Customers and Partnering

Any owner could be a candidate for partnering, but I've found that larger businesses that require ongoing work and maintenance contracts tend to be more receptive to the partnering concept. These would include businesses such as medical facilities, manufacturing companies, or retail chain stores as well as government or educational institutions. They're often companies or organizations that have been burnt in the past on competitive bid deals. Sometimes they'll even approach you with the idea — and you should be ready to jump at the opportunity. I think it's safe to say that all contractors like the security of an ongoing, long-term deal. When coupled with an assured profit that's fair to everyone, partnering is most always a good deal.

Introducing the Partnering Idea

When should you introduce the partnering idea if you think it'll work out well? The earlier the better. On a large project, you may want to schedule a meeting early on to discuss the partnering concept; you can make it sort of a "brain-storming" session with partnering as your end goal. One thing you can discover in a meeting like this is whether there's anything (or anyone) that poses a threat to the goodwill required to make partnering successful. In this meeting you can discuss everything of concern to the project, including scheduling, value engineering, personnel requirements, payment methods, conflict resolution, and so on. Don't worry about details yet, just start blocking out the process. Based on this initial meeting, you can cultivate ideas and methods of working together that'll eventually develop into a full partnering arrangement.

The Role of the Subcontractor and Supplier

When subs and suppliers agree to be involved in a partnering project, they've agreed to the pact. The general contractor will be sharing costs, markups, and invoices with the owner — and the subs and suppliers are expected to do the same. Granted, most communication comes through the general contractor, but the subs and suppliers are obligated to be open, honest, and supportive of the partnering plan. Since everyone is part of the team, there's no room for individual agendas and attitudes.

Every now and then, a subcontractor may be asked to break out the cost of individual items or separate labor numbers. It's no secret that they don't like doing this — no one does. For one thing, it's more work, but there's also a natural tendency to feel "overexposed." You know the feeling. With your numbers hanging out there, you know that it won't be long before everyone figures out what your profit line item is — and then sees if they can change it for you.

How can you make the subs and suppliers feel at ease in giving you this information? First, explain that the reason you need them to break out a number is because the owner asked you to break out your numbers. Second, let them know that you don't really care what their profit is. They've given you a professional, lump-sum quotation that includes the complete scope of the work, including punctual delivery and quality workmanship, and you're happy with it. And, that should include a profit. You're making a profit, so why shouldn't they!

Partnering and Design/Build or Negotiated Deals

The difference between partnering and design/build or negotiated deals is the *formal written agreement* promoting mutual respect, trust, honesty, good faith and open communication as the foundation of the working relationship. In other arrangements, the atmosphere of trust and goodwill may be generally *implied* at the beginning of the work, but it's left to wither on the vine as deadlines loom, mistakes bloom, and temperatures rise. With no *written* dedication to honesty and forthright behavior, finger-pointing and "revisionist history" (that all-too-common occurrence in which the past is remolded to meet present needs) can easily emerge to spoil relationships. With a written agreement, you have something tangible to help remind all parties of their commitment.

I know that this agreement is just a piece of paper — but pride in the fact that you've given your *word* is a remarkable thing. People like being perceived as ethical and honorable. We all remember, at least from old movies, when a man's word was his bond. The partnering agreement reinforces old-fashioned values and, I think, our natural inclination to want to live and work in an atmosphere of honesty and integrity. It's a tool that turns loyalty and commitment into something every bit as real as a circular saw.

Can You Become Too Chummy?

Familiarity may indeed breed contempt — and it's no different in construction, even if you're working with a partnering arrangement. Besides the language outlining the partnering agreement, your contract still needs to have the responsibilities of each party clearly defined. Everyone should be legally bound by written contract. The difference between this type of contract and one that doesn't include the partnering agreement is the establishment of a positive, productive working atmosphere within the written contract. The covenant to "get along" isn't just implied. Breaking any part of the pact is no more acceptable to the parties involved than leaving out a structural element of the building or not providing a specified material. Working well together becomes a critical element of the construction process.

Conclusion

Partnering and design/build concepts work because they're a better approach to construction for *all* the participants, *not* because they're the latest trend or because I say so. It's a let's-get-back-to-basics way of doing business. Designed to encourage common sense and healthy interaction, partnering addresses and reinforces aspects in all of us that too often go untested in today's business world — *honesty, integrity, and commitment.*

For More Information

If you'd like more information on design/build or partnering, you can find it at your library or you can access information on the Internet.

Contact the Design Build Institute of America (DBIA) directly at:

The Design Build Institute of America
1010 Massachusetts Avenue, NW, Suite 350
Washington, DC 20001
Phone: (202) 682-0110
http://www.dbia.org

You can access Design/Build Business Online for articles on design/build projects. They also have a chat room. They are on the Web at:

http://www.yourworld.net/dbb-online/dbb.htm

To find information on partnering on the Internet, go to:

http://construction.conncoll.edu/partnering/index.html

Gather the information. Consider the concept. Then give it a try when a suitable partner turns up. You may never want to go back to doing business the old adversarial way again.

Chapter 9

The Submittal and Shop Drawing Process

The *submittal and shop drawing process* is something that you most likely didn't encounter in residential work. Sure, you've delivered samples or gathered technical literature for your customers to view and approve, but it wasn't a formal process. This changes when you go into commercial construction. The submittal and shop drawing process, supervised by the architect, is both highly structured and mandatory. It's as much a requirement of the commercial construction contract as the construction itself.

The shop drawing and submittal process is the architectural/engineering firm's way of guaranteeing that the products, procedures, and applications called out in the original plans and specifications for the project are the *same ones* that end up being used in the actual project construction. The process ensures that not only the general contractor, but also the subcontractors and suppliers, comply with the established level of quality. The package *proves* that the products specified in the plans are the products that will be used and/or installed.

In addition, the process acts as a vehicle to flush out any details of the project that are flawed or weren't completely developed during the design and bidding stages. The shop drawings and technical data that make up the submittal packages are much more detailed than most plans and specifications. It's not uncommon for design problems to show up at this stage, such as a light fixture or HVAC duct that doesn't quite fit into a too-shallow ceiling cavity. In this way, the process acts as a check and balance system for the architects as well.

The Process

After the job has been awarded to a general contractor, the submittal process begins. It proceeds in much the same manner whether the job is a design/build or competitive bid project.

The first step in the process requires the subcontractors and suppliers to submit their packages to the general contractor. Generally, the first page or so of the CSI Division that pertains to their particular scope of work will spell out each trade or supplier's submittal requirements. Everyone should also read Division 1, General Requirements, in the manual to find additional submittal directions and responsibilities. For instance, a mason may be asked to provide mortar colors and brick samples for review. He may even be asked to construct a mock-up or small sample of the proposed masonry work to be reviewed and approved by the architect and owner before the actual full-blown construction begins. General Requirements is also where they'll find how many sets of submittal packages must be sent in. It's not uncommon to be required to provide six, seven or even eight sets of the same submittal for review.

The general contractor reviews the material, and if he's satisfied with the submittal, he stamps it "*approved*" or "*reviewed*" and forwards it to the architectural/engineering firm. They in turn review the same package. They can accept or reject it, request changes, or offer further direction regarding the package. If they accept the package, the architectural/engineering firm also gives it their stamp of approval, and then returns it to the general contractor. The general contractor relays it back to the subcontractor or supplier for processing. Figure 9-1 shows a flow chart of the process.

You'll note that there are a lot of people involved in this process — and it *does* take time. While it may seem awkward, it's a reliable method of keeping all the parties informed on the status of the submittals, including any changes that are made along the way. When you consider the one to three weeks it takes to relay the submittal package back and forth added to the time spent *creating* the package (often many weeks), you can see that this can be a very lengthy process.

This is bad news for the project manager who's trying to push a construction schedule along. Most suppliers won't order the material for a project until their submittal has been returned marked "approved." Lead times for products can range from weeks to months, and when coupled with the time required to complete the submittal process, it can be virtually impossible to arrange for special order materials to arrive in the first one or two months of the project.

1. Project is bid.

2. Project is awarded to a prime contractor.

3. Prime contractor awards contracts to subs and suppliers.

4. Subcontractors and suppliers prepare and send submittal packages and shop drawings to the prime contractor.

    4a. Prime contractor disapproves the package and returns it to subs and/or suppliers for corrections and resubmittal. Process returns to step 4.

5. Prime contractor approves the package and forwards it to the architect/engineering firm.

6. Architect/engineer receives package and either approves or disapproves it.

    6a. If architect disapproves package, it is returned to the general contractor for resubmittal. Process returns to step 4.

7. Architect/engineer approves the package and returns it to the prime contractor, who in turn, returns it to the subs and suppliers for processing.

8. Subcontractors and suppliers order materials and proceed with their contract obligations.

Figure 9-1
Submittal process flow chart

This can pose a serious problem for the project manager who needs to have hollow-metal door frames on the job site by the time the mason begins the lower-level block walls (usually about four to five weeks into the job). In a case like this, the project manager may be forced to call the hollow-metal door supplier and strike a deal guaranteeing that the general contractor will accept the responsibility for approving the product. That way, the doors can be delivered in the first month and the project can move ahead on schedule. This is a huge risk for the general contractor. All he can do is hope that the architect doesn't change the specifications during review of the package. If the product (in this case, the hollow-metal doors) is rejected or changed by the architect, the general contractor has to pull out the frames he's had installed and put in the new ones — usually at his own expense. Occasionally, however, you may find an architect who'll work with you on the extra cost if this happens.

All this goes on before the submittal package is accepted. *And many are not!* If a package is rejected, the submittals are returned via the same route, and the whole process starts all over again. A rejected submittal will often have comments from the architect explaining the reason for the denial. The submitter must fix the package and then resubmit it.

Many times a general contractor will take a leap of faith and proceed. You tend to know if a package will be approved or not, especially if you've worked with the architect, suppliers and subcontractors in the past. It's a risky way to do business, but if you're going to meet those project deadlines, you're tempted to take a running start on the project before "all the i's are dotted and the t's are crossed."

Tracking Submittals

The project manager keeps track of all the submittals for the project to make sure everyone complies — and that they're punctual in their delivery. He may use a submittal log like the one shown in Figure 9-2 for this purpose. He usually logs in four dates:

1. the date the submittal was requested from the sub/supplier

2. the date the sub/supplier delivered the package

3. the date the general contractor delivered that package to the architect

4. the date the architectural/engineering firm returned the package to the general contractor for distribution to the sub or supplier

If a subcontractor or supplier takes too long to deliver their submittal package, the contractor may send out a deficiency notice (Figure 9-3) to let them know they need to pick up the pace.

Types of Submittals

Submittal packages come in many shapes and sizes. We'll look at a few of the most common submittals required.

Samples

Product suppliers often need to provide an actual sample of the product to be used in the final construction. If color or other aesthetic choices have to be made, several sample packages may be requested so all the people involved in the decision will have an opportunity to review them.

Product Specifications (Technical Data)

Very often, submittal packages are just a matter of gathering together the pertinent technical information on the product to be used. This would include application and installation procedures, as well as warranty information. This literature is available through manufacturers, suppliers, wholesalers, or may even be located through *Sweet's Architectural Record* or similar references. It's common to submit four to eight sets of the package. Most of the time, if samples are requested, one is enough — but it's always best to check with the architect to be sure.

If "spec" or "cut" sheets are submitted, highlight the exact product to be used on each page to prevent confusion. It's useless to submit a catalog page full of similar, but different, products (with the one to be used among them) and expect that to pass. Employ a little common sense and cut and paste a custom job sheet from the catalog.

Submittal/shop drawing log for _____

ABC Contractors

1234 Hammertacker Way, Lienwaiver, WI 53555 • Office (608) 555-1234, Fax (608) 555-1235

Project name _____

CSI#	Firm name & contact person	Submittal/shop drawing description	Date of award to sub/ supplier	Date submittal/ shop drwg. received	Date forwarded to A/E	Date approved and returned by A/E	Date returned to the sub or supplier	Projected date product or service deliv'd to site	Actual date product or service deliv'd to site

File: SUBMTLOG

Figure 9-2
Submittal/shop drawing log

ABC Contractors

1234 Hammertacker Way, Lienwaiver, WI 53555 • Office (608) 555-1234, Fax (608) 555-1235

Deficiency Notice

_____
(Date)

To: _____   Project Name: _____

        _____   Project Number: _____

        _____   Owner: _____

Attn: _____   Contract for: _____

As of the above date, we are lacking the following material required as part of your agreement with us for the above project. Please submit the items checked below to our office as soon as possible in order to avoid any delay in project schedule and/or your payment:

❏ Signed subcontractor agreement   ❏ Purchase order (or PO acknowledgement)

❏ Certificate of insurance   ❏ Submittal/shop drawings/mock-up

❏ Schedule of values/cost breakdown   ❏ Lien waiver for (period): _____

❏ As-built drawings   ❏ Invoice for (period): _____

❏ Warranty information   ❏ Samples/product information

❏ Color charts   ❏ Other closeout documents: _____

❏ Manpower breakdown/project schedule   _____

Note: _____

_____

_____

_____

_____

Signed _____

File: DEFNOTE

Figure 9-3
Deficiency notice

Shop Drawings

Sometimes (depending on the discipline or trade), you'll have to provide more exacting and customized information than a standard submittal package. In this case, shop drawings may be requested. You'll probably have to submit this additional detailing for roof trusses, structural steel, hollow metal and similar items, as well as specific trade drawings like electrical schematics and plumbing isometrics. The drawings are generally bluelines (similar to regular architectural working drawings), and include cross sections, fine details, structural data and backup, and layout information. Make sure they're drawn to a specific scale, and compile several sets so they're available to all parties for review. Keep them available later as the project progresses.

> *"The review process not only lets the architect check for compliance to the specifications, it also gives him a second chance to detail and/or modify the technical aspects of the project . . ."*

When shop drawings are required, the cost to the subcontractor or supplier goes up. Gathering catalog cut sheets may only take a little time, and that can usually be absorbed in the day-to-day cost of doing business. But shop drawings require much more time, money, and effort. Companies that create shop drawings often employ a full- or part-time draftsman to churn out the work. You can sub out the drawings to a third party, but that doesn't offer as much control and reliability as in-house drawings.

No matter how the drawings are produced, they'll cost money — and that needs to be accounted for in the bid. This cost can sometimes be subsidized or eliminated by using supporting companies, manufacturers or suppliers who employ their own engineering departments and offer design and/or shop drawing services to their customers as part of their package. Manufacturers of roof trusses and pre-engineered buildings, for example, often offer this type of service.

Mock-ups

A mock-up is a small example of what will be constructed on the actual project. Whether it's a section of an exterior masonry wall or a panel of glazed curtainwall, all of the detail and quality required in the finished construction must be included in the model. A mock-up is generally built right on the project site so the owners and design professionals can use it to verify the finishes, workmanship, and overall blending of the detail into the project. The cost to construct a mock-up can be substantial. Subcontractors finding this requirement in the specifications have to include the cost of constructing the mock-up in their bids for the general contractor.

Why Have a Review Process?

The creation of a commercial building project, with all its peripheral work, is extremely complex. Even the finest architectural/engineering firm wouldn't be expected to detail every single aspect of a project's construction — nor would an owner want to pay them to do it. It simply wouldn't be economical. So, much of the specific detail work — the nuts and bolts of construction — falls on the shoulders of the general contractor, the subcontractors, and the suppliers who are awarded the work. The review process not only lets the architect check for compliance to the specifications, it also gives him a second chance to detail and/or modify the technical aspects of the project before construction begins.

A Disturbing Trend

Giving the individual contractors and suppliers more input into the design and construction process can be a positive thing, since it often results in value engineering and speedier schedules for the owner. After all, who knows more about the latest trends, cost-saving alternatives, and construction methods than the professionals who perform the work on a daily basis? However, recently this shift in responsibility has led to a disturbing trend among designers. Designers aren't just allowing contractors more say in the work being done, they're gradually *transferring* the design responsibility for certain aspects of a project — and the liability for those designs — to the contractors.

I work with a number of architects who no longer even *attempt* to design the mechanical or electrical portions of the project. Instead, they opt to simply supply the architectural layout to the bidding contractors to design/build their own electrical or mechanical designs. Then those contractors supply an estimate for the work they've conceived.

This creates several problems:

▌ First, it opens up the bidding process to an apples-to-oranges scenario. Since each sub will come up with a slightly different design with a different cost estimate, it's more difficult for the general contractor, architect, or owner to really compare the proposals they receive. While they may be similar, they're probably not identical.

▌ The second problem is that since the architect doesn't design these systems, he often waives any responsibility for the review process, making the contractor entirely liable. The contractor can pass the responsibility on to his subcontractor, but both he and the architect have approved the design provided by the sub so the ultimate responsibility remains murky.

▌ Finally, without a check on the quality of the design, the door is open for less-reputable or less-experienced contractors to provide designs that are low in cost, but possibly substandard.

Unfortunately, this design-shifting trend isn't likely to end. After all, the architects are the ones who create the bid packages that contractors have to rely on to earn a living. They draw the plans, write the specifications, and create the wording in the forms and documents used for most commercial construction. These packages are set before us, and we're often faced with a *take-it-or-leave-it* situation. And because of competitive pressure, we usually end up taking it.

Checks and Balances

But contractors aren't saints either. The architectural review process also operates as a check on the people and the money. As long as one chases the other, there'll always be a few of us who'll try to take advantage. You might be tempted to slip in something a little cheaper, that's slightly lower in quality or a littler easier to install, in the hope that no one will notice and you can pick up a few dollars. As an industry, it's not something we're particularly proud of, but it does happen. The review process itself acts as a deterrent to those who would consider taking shortcuts.

Of course, not all problems caught in the review process are deliberate attempts to defraud. More often than not, problems with compliance to specifications are genuine misunderstandings between the contractor/supplier and the architect regarding the meaning of the specs. This is particularly likely to occur in cases where products or services are submitted by a contractor under an "or equal" clause in the specification manual. The "or equal" clause commonly allows for products of equal or better quality to be offered up by the contractor when quoting a project. The architect almost always still has final approval, and the contractor may have to provide supporting data to prove that the item submitted is equal or better. Although it can lead to problems, the "or equal" clause does provide a channel for an open dialog, and that may result in an enhancement of the project.

Final Thoughts

Depending on the job and the architect, there can be variations to any of the procedures we've discussed. I've had jobs where I wasn't required to submit anything for approval. The architect and owner basically trusted us. Then again, I've had projects where several mock-ups had to be constructed, reviewed, and approved before actual field construction could go further. When in doubt about what to submit, *read the specification*. If you don't find what you're looking for there, ask the architect. The object is to eliminate misunderstandings in the early phases of a job, when they won't cost you nearly as much money as they will once you're in the field.

Chapter 10

Subcontractor and Supplier Selection

One of the most important (yet surprisingly underrated) skills that you need to be a successful commercial contractor is the ability to select professional, efficient and cost-competitive subcontractors and suppliers to help you perform your work. Their value to you can't be understated. Even if you've done everything in your power to prepare for the upcoming project, if your subs and suppliers fail to deliver — your project will fail.

What this means is that the commercial construction process is a *social exercise* as well as a technical one. Sometimes you'll feel more like a social worker than a construction contractor, but that's all just part of the commercial contracting experience. Everyone knows that commercial construction isn't something you do alone. It involves a team, composed of different companies and individuals, who bring a variety of talents, abilities, temperaments, and backgrounds to the construction project.

My philosophy in writing this book is to show you all of the positives *and* the negatives involved in the work, so you can use that information to come to your own conclusions about what's important or relevant. That way, with the whole picture before you, you'll be able to plan your own strategy to deal with potential obstacles.

Here's one of those stumbling blocks. The pool of subcontractors and suppliers that you'll have to deal with is often a less-than-perfect mix. Like everything else in life, there's good and bad. It's almost a certainty that at some point you'll find yourself working side-by-side with individuals who, under better circumstances, you wouldn't want to deal with. That's natural. They may feel the same about you. Over the years I've worked with subs and suppliers that I personally couldn't stand, and there were people that I considered friends who didn't ever work with me.

Do you think that seems odd or cold? Well, it isn't if you understand how the commercial general contractor selects subcontractors and suppliers for his projects. Selection is a business decision, and in commercial construction your business is to deliver the highest quality of construction for the absolute lowest dollar amount possible. It's that simple.

With economic pressures constantly looming, it doesn't take long for even the most soft-hearted contractor to put aside personal feelings and use analytical evaluation to select the best subcontractors and suppliers for the job that needs to be done. You'll find that what may appear to be a relatively straightforward arrangement is, in fact, an exercise of immense responsibility and concern to every contractor.

The Selection Process

There are a lot of fairy tales floating around about how contractors select subs and suppliers for a particular project. The biggest tale, of course, is the wide-held belief that the selections automatically go to the lowest subcontractor bid and supplier price. I can tell you from experience that this isn't so. The never-ending quest for the perfect bargain will always be a motivating factor — but it's not the only factor. There are other considerations as well. But before we look at these other considerations, you need to understand how a commercial general contractor evaluates the sub and supplier proposals he receives.

When the commercial contractor is putting his estimate together, he'll usually get a number of estimates from subs in a particular trade that are very close in cost. I'm not talking about a 5 or 10 percent difference, but rather a difference of only 1/2 to 1 percent. For example, on a recent competitive bid proposal I had five bids within $1,000 of each other on a $115,000 HVAC installation. Though it's rare for so many sub bids to come in this close, it does happen.

So what did I do? I viewed all of these bids as being the same number. This is something you have to do, particularly when there isn't time to adequately define the scope of work for each proposal during a hectic bid period. And though the bids were not so close in the other divisions of work, I also grouped them competitively at the low mark. In most cases, you'll find that the variation in bids just isn't that great.

Contrary to popular belief, you seldom get that mythological *"low"* number that many subs and suppliers assume exists. As a matter of fact, when you do get a bid that seems inordinately low, it tends to create suspicion. Why? Because it would be bad business (and bad ethics) to knowingly award a job to someone whose bid is so low he can't possibly make a profit on the job. Besides the apathy that would almost surely plague his performance once he discovered his mistake, there's also the very real possibility that the sub would default on the job completely. That would leave the contractor, and a very nervous owner, with a gaping hole to fill in the construction schedule.

Sub and Supplier Considerations

Assuming you get similar sub and supplier bids, which eliminates the cost consideration, you must look at other selection criteria to decide who's best

for your project. Figures 10-1 and 10-2 are comparison sheets I use to help me organize pertinent information on the subs and suppliers bidding my jobs. These sheets aren't overly detailed, but contain enough general business and bidding information to help me make my choices.

There are several items that you should consider — beginning with experience.

Experience with the Type of Work Being Awarded

Let's assume your project is a new wing on a hospital, and the hospital is a repeat customer of yours. Because you've dealt successfully with them before, you have an incentive to stay in their good graces. Before you hire a sub or supplier for this job, you'll want to be sure that they understand, and will comply with, the unique considerations of hospital construction. There's the strict work documentation (stricter than most private projects), interim life safety training, code response for hospital emergencies, extra concern for cleanliness, and the quiet operation required for work in a hospital environment. All of these special challenges translate into additional responsibility — and extra manhours. The sub or supplier who has worked under these conditions before will be a step ahead of those who haven't.

Manpower

Providing adequate manpower doesn't simply mean that the subcontractors will bring lots of workers to your job site. It means that they employ full-time, trained, experienced craftsmen, and they have enough employees to complete the work. It also means that their workers will show up every day on time and be prepared to work. You don't want subs who go running to the temp agency and make up crews on the fly whenever there's a pinch for help. Temporary workers are just that: *temporary*. They're a stop-gap solution that's almost always detrimental to the project's overall efficiency. It may benefit the subcontractor's bottom line, but from your point of view, there's never enough money in the budget to have unproven workers learning their trade on your project's dime!

ABC Contractors

1234 Hammertacker Way, Lienwaiver, WI 53555 • Office (608) 555-1234, Fax (608) 555-1235

Subcontractor Comparison Sheet

Project Name: _____

No.	Description	*Subcontractor	*Subcontractor	*Subcontractor	*Subcontractor
1	Insurance limits including umbrella				
2	Bond rate & limit (if applicable)				
3	Age of company / annual volume				
4	Defendant in any litigation in last (3) years				
5	Acknowledged receipt of drawings				
6	All products and applications according to specification				
7	Full-time superintendent provided				
8	Number of personnel in office/field				
9	All applicable sales and use taxes included in bid (rate %)				
10	Number of days required to mobilize from verbal notice to proceed				
11	Daily cleanup included in bid (rubbish removal into GC dumpster)				
12	Completed work guarantee period				
13	All closeout and warranty procedures included in bid				
14	Labor rates for additional work (through) * enter date				
	Indicate % increase (if any) for labor after * enter date				
	Foreman: standard time / overtime				
	Journeyman: standard time / overtime				
	Laborer: standard time / overtime				
15	Can meet owner project schedule for substantial completion by:				
16	Will work overtime/weekends to maintain schedule at no additional cost				
17	Will attend weekly site subcontractor job meetings				
18	All hoisting, cranes, elevators, etc. included in bid				
19	Long-lead items and delivery dates: (use separate sheet if necessary)				
20	Work complies with all codes having jurisdiction				
21	Bid price guaranteed to date:				
22	Base bid:				
	Alternate number 1 (description):				
	Alternate number 2 (description):				
	Alternate number 3 (description):				
23	Total bid (project):				

File: SUBCOMPR

Figure 10-1

Subcontractor comparison sheet

ABC Contractors

1234 Hammertacker Way, Lienwaiver, WI 53555 • Office (608) 555-1234, Fax (608) 555-1235

Supplier Comparison Sheet

Project Name: _____

No.	Description	Supplier	Supplier	Supplier	Supplier	Supplier
1	Age of company / annual volume					
2	Defendant in any litigation in last (3) years					
3	Acknowledged receipt of drawings					
4	All taxes included (rate %)					
5	Number days for shop drawing / submittal delivery					
6	Delivery dates (incl. shop drawing/submittal period):					
	Product: Date req'd/delv'd:					
	Product: Date req'd/delv'd:					
	Product: Date req'd/delv'd:					
	Product: Date req'd/delv'd:					
	Product: Date req'd/delv'd:					
7	Warranty periods					
	Product: Warranty period:					
	Product: Warranty period:					
	Product: Warranty period:					
	Product: Warranty period:					
	Product: Warranty period:					
8	Will attend weekly/biweekly site subcontractor job meetings					
9	Price includes all hoisting, elevators, crane, etc. to place material on site					
10	Prod. complies with specifications and all codes having jurisdiction					
11	Bid price guaranteed to date:					
12	Base bid:					
	Alternate number 1 (Description):					
	Alternate number 2 (Description):					
	Alternate number 3 (Description):					
	Alternate number 4 (Description):					
13	Total bid (project):					

File: SUPCOMPR

Figure 10-2

Supplier comparison sheet

Temporary workers also affect worker morale on the job. Seasoned professional tradesmen take their skills very seriously. The fact that a less-skilled workman can simply walk in off the street and work beside them undermines their self-esteem. I've observed over the years that when temporary workers are brought in to supplement crew size, the resulting lack of experience wears on the company's veteran crew members. If they have to work harder to pick up the slack, their motivation and their morale plummet — and so does your profit.

Tools and Equipment

Obviously, the subcontractors and suppliers you select need to be properly equipped to perform the work they're contracting to do. But what exactly do you mean by *properly equipped*? It depends on the job they'll be doing. Does the mason you're considering have the proper staging, scaffolding, and lifts needed to complete the high work required of the project? Does the concrete sub have the curved forms for pouring the water tank? Does the plumber have core-drilling equipment to drill through the existing foundation walls? These are variables that will change from job to job. Not every sub or supplier will be equipped to perform every type of job. The logic in your selection should be pretty simple. If the scope of work involves the prolonged use of a particular type of equipment, a special skill or a particular type of material, the company that has that skill or equipment or material will have the competitive advantage when it comes to earning the job.

Financial Strength

General contractors don't inquire about a subcontractor's or supplier's financial situation simply because they're nosy. If a subcontractor or supplier goes belly-up midway through a job, it can have devastating consequences on the entire project. First, of course, because you have to quickly replace them, or your schedule is delayed. But there are other consequences as well, like the trickle-down panic that finds its way to the project owner, who may suddenly be an unwilling participant in the defaulted company's financial and/or legal misadventures.

Even if the owner doesn't get dragged into that company's dilemma, there's still the problem of bringing a new sub or supplier on board midway in the game. This almost always involves increased costs to the general contractor. In most cases you'll offer the contract to the next highest original bidder that you felt was qualified for the job. Keep in mind that a replacement subcontractor probably won't be just sitting around waiting to step into this new role. They'll almost certainly need some time (days, perhaps even weeks) to make the transition onto this new project. And what about the submittal and shop drawing process? Will you need to go through it again with the new company?

A mid-job replacement process is always messy. You'll have to:

1. Determine and close out the defaulting contractor's scope of work, which is never, ever clean and without disagreement.

2. Select and orient a new subcontractor into a job that's partially complete, which can take time.

3. Deal with the negative mood that inevitably falls over a project when a member of the team goes under.

And what about the warranty on the work? Who's going to warrant the defaulting contractor's work now that there's no company to back it up? Messy? You bet! That's why financial strength should always be an important consideration in your selection.

Proving financial stability doesn't mean that the sub or supplier must show you a sparkling, four-star financial report card. But they do have to be healthy enough to handle the work in question. The importance of the sub or supplier's financial situation is proportionate to the size and scope of their role in the project. You should be far more concerned about the financial strength of the mechanical subcontractor who has a $220,000 contract than the supplier whose contract totals $1,500.

Design/Build Capabilities

In recent years, it's become more and more important for subcontractors to have the capacity to provide in-house design/build services. They're frequently asked to do some or all of the design for their particular scope of work because architectural firms are providing less technical trade information than they used to. Market economics and competition have forced many design firms to spend less time and resources on initial design, especially for mechanical and electrical systems. They pass on these responsibilities, in the form of design/build, to the individual trades who subcontract for the jobs.

> *". . . . subcontractors who recognize the need for design/build services . . . . have an advantage over those who don't."*

Like it or not, since it's economically driven, this shift in design responsibility is probably around to stay. The subcontractors who recognize the need for design/build services and who build an effective design/build program have an advantage over those who don't. However, they shouldn't just give this work away — they should charge for it! The architectural firms certainly do! Subcontracting firms should think of it as another service that they can offer.

Each company should develop a design/build method that suits their individual needs. They may want to sub their design/build work out to third-party designers or engineering firms until they're in a financial position to bring in a full-time designer on the payroll. Whatever route they decide to take, offering design/build services not only makes a company more valuable to the general contractors and owners putting out the bids, but it can grow into a separate profit center for them. This holds true for general contractors who need to provide design/build services.

Quality of Work

No one wins when work is performed shoddily and carelessly. Since poor workmanship is never acceptable, tradesmen who perform poorly will usually be asked to correct or replace the offending work at no additional cost to the owner. If you've been in the construction business long, you know that redoing substandard work often ends up costing a lot more than the direct cost of replacing the work. There are trickle-down effects of poor workmanship that also harm the project.

The job schedule is almost always interrupted or altered while the work is being corrected — and that affects other subcontractors and suppliers. There's also the administrative time required to spot the error and order a remedy. And the owner will almost certainly become anxious about the general quality of the work being performed by all of the subs and suppliers on the project. It only takes one substandard installation to create mistrust between the owner and contractor. If it isn't smoothed over quickly, it can poison the entire atmosphere on the job.

General Reliability

The more reliable a subcontractor is, the less important all the other considerations become. Nothing beats a subcontractor or supplier who comes in on time, keeps his promises, performs his work with care, doesn't ask for petty change orders, and fulfills his obligations according to the original plans and specifications. That seems simple, doesn't it?

Unfortunately, however, there are way too few subcontractors and suppliers out there who are consistently that reliable. I absolutely guarantee that if subcontractors and suppliers would always deliver a quality product at a fair price, show up when they're needed, and deal honestly and in a straightforward manner with owners and other contractors, they'd *never* have to bid the lowest price to get the job!

Cost Reliability

There's another type of reliability that's an important part of the bid consideration. It's what I call *cost reliability*. This is one of my pet peeves, so I put a lot of weight on cost reliability when I let out a subcontract. Before I consider a sub for a job, I ask myself this question: *Does this subcontractor have a history of initially bidding low and then hitting up the general and the owner for every change order that they can dredge up?*

Obviously, if I know from past experience that I'm going to have to add another $4,000 or $5,000 to a sub's original bid to cover the cost of items that weren't included, then his low price isn't really low. You don't want to let a sub do that to you more than once. I'm much better off going with the next highest subcontractor — or one that I know won't play these games.

Speed of Work and History of Delays

The speed at which a subcontractor performs his work is another very real factor in getting awards. No one wants to work with someone who drags his feet getting the work done. A delay in a projected schedule almost always translates into lost profit for everyone on the project. When you get off schedule, you suffer lost manhours, possible liquidated damages, declining morale, and problems with the owner. Remember, delays cost the owner too. They may have to pay interest penalties, or they may lose their prospective tenant's first month's rent — or worse, they may lose the tenant altogether if their lease is based on a rigid move-in date. That's why commercial construction schedules are taken so much more seriously than they are in residential construction. There's a lot more money at stake.

But *speed* isn't just performing the work quickly. Most tradesmen work at about the same pace. Delays come less from the individual workers, and more as the result of a subcontractor's inexperience or inability to plan, stage, schedule and manage their part of the job in a timely manner.

Professionalism in Handling Architectural Protocol

This probably doesn't rank up there with some of the other considerations we've discussed, but I still feel handling architectural protocol is important. A sub or supplier should have the ability to deliver complete and professional shop drawings, submit-

tals, mock-ups, and anything else required under normal architectural procedure. They should also supply testing and certifications without having to be asked, and perform warranty and closeout procedures according to the specification. In short, they should live up to the administrative requirements spelled out in specification manual for subcontractors and suppliers.

Even though all the participants in commercial construction projects have these responsibilities, the vast majority of subs and suppliers don't follow through on these items as they should. This is another way a subcontractor or supplier can stand out. If they become knowledgeable and proficient in architectural protocol, that'll make them a more valuable team member. Knowing a sub or supplier will follow through on all their responsibilities means that the general contractor has one less thing to worry about — and that's an automatic advantage in the bid process.

The Final Consideration

There are other considerations as well — and these are as varied as the personalities of the people involved. In fact, it's possible that a sub or supplier may never know exactly why they were or weren't chosen for any particular job. Geographical distances, civic or business affiliations, and "local" advantages may also enter into the decision. But one thing is certain: the decision will always be largely based on the subcontractor's or supplier's integrity, reliability, and skill.

Purchase Orders and Subcontractor Agreements

Once the players are chosen, all that's left is to award the work in writing. Most often, this is accomplished through the use of *purchase orders* and *subcontractor agreements*. Some builders still award work with a promise and a handshake, but I think most of you have been around long enough to understand the importance of putting agreements on paper. You simply never know when a relationship will head south. Considering the contract amounts at stake in our industry, you need to protect yourself and your company.

A purchase order is generally simpler than a subcontract agreement. They're used for purchasing material, such as lumber or millwork items like windows and doors. They usually include relevant information about the product being ordered, such as the general description, brand, quantity, shipping terms, payment terms and cost. An example of a completed subcontractor purchase order is shown in Figure 10-3 and a supplier purchase order in Figure 10-4. You can use the same form as a fax purchase order by simply including a cover letter.

Subcontract agreements are more detailed, including more information about the actual performance of the labor itself. Figure 10-5 is a simple subcontractor agreement. You can get subcontractor agreements through construction organizations like the AIA (American Institute of Architects), the AGC (Associated General Contractors), or the ABC (Associated Builders & Contractors).

The "Buy-Out"

The subcontractor and supplier selection process is part of the commercial construction "buy-out period," commonly called just the *buy-out*. It's during this time, immediately after the job has been awarded to the general contractor, when he makes the commitment to — or buys — the subcontractor and supplier proposals that he'll use on the project. This is a serious time for the ones making the selections (normally the project manager or the estimator, or perhaps the two working together), perhaps even more serious than the bid period itself. Why? Because this is the last opportunity the general contractor has to thoroughly evaluate and compare each bidder's scope of work against the dollar value of the bid proposal. Or better put, *this is the last chance the general contractor has to "lock in" his costs.*

Here's another thing to note. It's not uncommon for the contractor to discover that some of the quotes he had initially thought were low and *complete* are actually lacking in scope and/or specification. This happens because there's simply *never* enough time, even with the finest construction estimators, to weed out every error and inconsistency that exists in the sub and supplier quotations before the final bid. It's only after the dust has settled, and there's time to analyze each quote, that a thorough investigation is possible.

And plenty of problems are exposed at this time. For example, you may find that the low EIFS contractor didn't bid the specified stucco, nor did he note on his bid that he was using a voluntary-alternate. Or, perhaps the HVAC sub didn't include gas piping because he thought the plumber was going to provide it. And then there's the painter; he left an entire room out of his bid! Just when you think you have everything sorted out, the low masonry contractor calls. Something has come up and he may not be able to do the job.

Unfortunately, situations like these are all too common, and they can put the general contractor in an awkward financial position. Of course, he has to do what's best for his company, and that means occasionally holding a sub or supplier's feet to the fire — or in this case, making them honor their bid.

This is why I earlier stressed the importance of the *social* aspect of the construction business. As a contractor and businessperson, you'll occasionally find yourself doing some hard negotiating, cajoling, and convincing to get your chosen team members to fall in line. It's part of the game. But, there's a catch. While doing all this hard negotiation, *you can't alienate or offend the other party.* Why? Because not only do you have to work with them on this project, but the odds are pretty good that you'll be dealing with them again and again. Think about it. If you stay in the same local market, you may work with only three to five different masons in your entire building career. There may not be dozens to choose from, so you'd better try to get along with everyone.

However, in spite of your best efforts, it's likely that some of your relationships will go sour. Jobs go bad and disagreements will spring up that can alienate you from one or more subs or suppliers. This alienation can sometimes last for years. But even

Purchase Order

From:	**_ABC Contractors_** *1234 Hammertacker Way* *Lienwaiver, WI 53555* *Office (608) 555-1234, Fax (608) 555-1235*	**Purchase order number:**	99-245
		Vendor Code:	DROP099
		Job name:	Baker Office Building

To: Dropcloth Painting, Inc. 5555 Main Street Latex Way, WI 55555 Attn: Pete Wilson	Ship to: Baker Office Building 3333 Anaconda Way Lake Geneva, WI 55555

P.O. Date	FOB	Ship via:	Per your proposal #:
9/4/00	Jobsite	Your truck	67468

Buyer	Freight	Remarks	Tax	Req'd Date
SSS	Included		Included	Per schedule

Quantity	Items #	Description	Unit	Ext. Cost
		Furnish all material, labor, and equipment necessary to perform interior painting work for the project listed above according to sheets A000, A109, A110, A111, A112, A119, A119.1, A120, A121, A124, A130, A131, A132, A133, A140, A141, A141.1, A142, A143, A144, & A145 of the project plans (including finish schedules) and the project specification (as it pertains to your scope of work) as created by T-Square Architecture & Design, Inc. dated 7-28-00 and also according to your proposal dated 8/8/00 and subsequent painting scope interviews conducted with ABC Contractors. Work includes painter's prep (as defined in specification), painting, staining, & finishing of all interior elements including millwork items as defined by the plans and specs, polychrome paint color schemes, faux finishes, special finishes, punchlist completion, and closeout & warranty procedures. Dipping/refinishing of doors as designated, mock-ups as necessary and minor patching & sanding. Dropcloth to include all necessary staging/scaffolding, dropcloths, protection of existing architecture, and equipment to perform their work. All work is to be completed no later than the dates set forth by the owner's representatives. PLEASE REFERENCE THE ABOVE P. O. NUMBER EACH TIME YOU INVOICE FOR THE ABOVE DESCRIBED WORK.	Total	$46,349.00

Authorized Signature: _____

File: PO

Figure 10-3

Subcontractor purchase order

Purchase Order

From:	**ABC Contractors** *1234 Hammertacker Way* *Lienwaiver, WI 53555* *Office (608) 555-1234, Fax (608) 555-1235*	Purchase order number:	00-245
		Vendor Code:	BFLC10
		Job name:	Baker Office Building

To:	Boardfeet Lumber Company P. O. Box 123 Janesburg, WI 55545 Attn: Dick Williams	Ship to:	Baker Office Building 3333 Anaconda Way Lake Geneva, WI 55555

P.O. Date	FOB	Ship via:		Per your proposal #:
9/4/00	Jobsite	Your truck		10368

Buyer	Freight	Remarks	Tax	Req'd Date
SSS	Included		Included	11/1/00

Quantity	Items #	Description	Unit	Ext. Cost
		Supply all wood doors according to the plans and specifications for the project above and your proposal dated 8-25-00, including:		
7 ea * *		3/0 x 6/8 - 1 3/4" Mahogany custom single-panel interior door - slab form (to match existing Driehaus Estate profile). No milling /prep /bore. Unfinished.	1,600.00 * *	$ 11,200.00 * *
10 ea		2/6 x 6/8 - 1 3/4" "Same"	1,410.00	$ 14,100.00
1 ea		3/0 x 7/0 - 1 3/4" "Same"	1,200.00	$ 1,200.00
2 ea		2/6 x 7/0 - 1 3/4" "Same"	1,500.00	$ 3,000.00
1 ea		3/6 x 7/0 - 1 3/4" "Same"	1,800.00	$ 1,800.00
1 ea		6/0 x 6/8 - 1 3/4" (Dbl) "Same"	3,050.00	$ 3,050.00
1 ea		6/0 x 7/0 - 1 3/4" (Dbl) "Same"	3,825.00	$ 3,825.00
1 ea		5/4 x 6/8 - 1 3/4" (Dbl) "Same"	2,850.00	$ 2,850.00
1 ea * *		2/6 x 6/8 - 1 3/4" paintable grade custom 6-panel interior door - slab form (to match existing Driehaus Estate profile). No milling /prep /bore. Unfinished.	1,000.00 * *	$ 1,000.00 * *
13 ea		2/6 x 7/0 - 1 3/4" "Same"	1,010.00	$ 13,130.00
4 ea		3/0 x 7/0 - 1 3/4" "Same"	1,050.00	$ 4,200.00
1 ea		4/0 x 7/0 - 1 3/4" "Same"	1,900.00	$ 1,900.00
1 ea		5/0 x 7/0 - 1 3/4" "Same"	1,920.00	$ 1,920.00
			Subtotal	$ 63,175.00
			Tax	$ 3,474.63
		PLEASE REFERENCE THE ABOVE P. O. NUMBER EACH TIME YOU INVOICE FOR THE ABOVE DESCRIBED WORK.	Total	$ 66,649.63

Authorized Signature: _____

File: PO

Figure 10-4
Supplier purchase order

Subcontract Agreement

This agreement is between Contractor:

Date: _____

_____

Owner: _____

_____

Project name: _____

_____

Project address: _____

and Subcontractor:

Insurance company & policy #: _____

Amount of insurance: _____

_____

Subcontractor federal ID #: _____

_____

Subcontractor license number: _____

_____

Description of work to be performed:

_____

_____

_____

_____

_____

_____

_____

_____

_____

_____

Payment schedule: _____   due on or before _____

_____   due on or before _____

_____   due on or before _____

_____   due on or before _____

_____

for the total price of $ _____

Work shall start on: _____   and shall be complete within _____ number of days.

Contractor's signature: Subcontractor's signature:

_____   _____

Date _____   Date: _____

File: SUBCONTA

From *Construction Forms & Contracts*, Craftsman Book Company

Figure 10-5
Subcontractor agreement

then, it's been my experience that this type of animosity tends to fade in time, and eventually the sub or supplier returns to the fold. I've been doing commercial construction long enough to see this phenomena occur not just once, but over and over with the same sub or supplier. Some people are just hard to work with (and they probably think it's me).

Being aware that this "circular game" is just part of the business will help you be a better negotiator and contractor. It doesn't pay to hold grudges or take occasional desertions by subs or suppliers too personally. Remember, you need them. They're your bread and butter, and probably one of the reasons why you remain competitive. So, while you're trying to be a tough and fiscally-responsible businessman, you also need to be a diplomat. Sometimes that's not easy. Even though you could *technically* make the sub who missed his bid "eat" the error, you may discover that if you want to use that sub again, the best approach is to try find a middle ground, one that you can both live with.

If an error turns out to be too much for either the sub/supplier or you, as general contractor, to absorb, then the problem is more serious. Remember, you've made a price commitment to the owner, and you don't want to have to go back on it unless you have no other choice. If that's the case, however, you may have to plead your case with the owner — and hope for understanding. Then, once the numbers are finalized, it's forgive, forget . . . and *full steam ahead!*

Chapter 11

The Profitable Job Site

There's another thing you should know about commercial construction: *your foreman is walking around with a toolbox full of your money*. And he's not the only one — so are your carpenters, cement finishers, even your own project manager! They all have your money hidden in their trucks and stuffed in their pockets. But don't panic; that's just part of doing their jobs.

You see, they don't actually have cash on them. What they do carry with them, however, is a dollar value for every minute they're on the job. And that dollar value is real and accountable, and directly impacts your profit line item — and the financial health of your company.

Making or Losing Money in the Field

Out on the job site, value comes in many different forms. First, there's the value associated with the actual amount of time that your worker is on site performing his task. This value is commonly referred to in *manhours*.

A manhour is a unit of measure for the amount of time that it takes a worker to perform his or her particular task or job. For instance, an electrical company estimator might assign $1^1/_2$ manhours for each duplex receptacle to be installed in a building. If there are 30 of these receptacles shown on the floor plan, the estimator simply multiplies the quantity (30) times the manhours (1.5 MH) and then multi-

plies that product again by his hourly billing rate for that trade. We'll use $40 per hour for our example:

$$30 \text{ units x } 1.5 \text{ MH} = 45 \text{ MH}$$

$$45 \text{ MH x } \$40 = \$1,800$$

This gives him a *direct labor cost* of $1,800 for installing all of the receptacles in the building. The estimator continues assigning manhour units to all the different work items (switches, light fixtures and so on) until he compiles his total direct labor cost for the job.

He then goes on to assign general requirements, overhead, and profit to those line items as he sees fit. General requirements, as you recall from Chapter 5, are those real costs that you assign to work performed in the field that aren't accounted for in direct labor and material costs or overhead. These would be items such as supervision, mobilization, trucking and freight charges, temporary utilities, and so on. It takes strict discipline and control by all of the workers in the field to insure that the general requirement costs don't get out of hand.

Productivity and Profit

Let's look at the relationship between *manhours and profit*. Suppose your sub has calculated that his carpenter will take 7 manhours to install each door frame, including the door and all the door hardware. Let's assume that your estimator is experienced and has done his homework, and he has arrived at the same manhour estimate using industry standards and your company's past historical records. In short, you're comfortable that this estimate is reasonably correct.

But once on site, the carpenter actually takes $8^1/2$ manhours per door instead of the estimated 7. *"Well,"* you might say, *"it's just a $1^1/2$ hour difference. We'll make it up."* Did I mention that there were 92 doors on this project? Or that you're committed to the owner for the original quoted (lump sum) price, and you can't return to ask for more money? You can see where this is going. You're watching a chunk of your profit march right out those slowly-installed doors! That's 92 doors times the 1.5 manhour per door underestimated: 138 additional manhours! At $34 per hour, you're looking at an additional carpentry cost of $4,692. It doesn't take a mathematical genius to see that you need to correct this problem — and quick!

But what do you do? Well, like most things in life, there's no single solution to controlling job site loses. Often you'll need to come up with a lot of little solutions. One of the commercial contractor's most effective weapons against job site loss is the ability of his superintendent and/or construction manager to recognize, address, and correct what I like to call "profit-depleting environments." Profit-depleting environments are those work situations that offer a fertile breeding ground for loss and waste at the job site. The trick, of course, is to know what to look for. Let's take a look at some profit pitfalls and examine remedies that you may want to apply to reduce their effect.

Communication

Blaming problems on poor communication may sound like a cliché, but you'll notice that I listed communication first. There's a good reason for that. In my opinion, lack of good communication, mainly between the office and the job site, is the largest single cause of lost profit in our business. As basic as it may sound, most mistakes in construction are the result of one person either not understanding anoth-

er person's instructions, or intentionally ignoring them. The truly frustrating thing about communication problems is that they're so easy to avoid with just a little effort and simple interaction.

Unfortunately, some people just don't try to communicate — especially stubborn, grizzled construction workers who don't like change. You may have to apply a little direct pressure to get things done your way. If you back it up with praise and periodic bouts of reinforcement, you'll get what you set out to achieve. Make it clear that good communication isn't an option; it's a mandatory part of everyone's job description! Make every worker on site responsible for the project. If there's a problem or conflict, let them know that you want to be told about it. Then make sure you provide a vehicle for open communication with you and/or your superintendent. If you let them know that you value their input (and make sure your superintendent reinforces this policy), you'll get their input either directly or through your superintendent. Usually he'll be the one who relays the information back and forth at his regular job progress meetings.

I've found from experience that the only way to accomplish consistent job site communication is through some type of *mandatory* interaction, such as regular on-site progress meetings. These meetings may or may not be an option. The original job specification manual may require you to hold them. But whether they're required or not, be sure to hold them on a regular basis and invite everyone, including the owner(s), the architect and engineers, and all your subs and suppliers. No one who's involved in the construction process should be exempt from these gatherings. For an average-size commercial construction job ($1 to $3 million), you'll probably want to schedule construction progress meetings at fixed intervals of at least every two weeks — and *demand* that all trades (or at least all the trades *currently on site)* be represented at the meetings. As the job progresses, and more details need to be hashed out, you may want to schedule a meeting every week, or even more frequently, if needed.

> *". . . . lack of good communication, mainly between the office and the job site, is the largest single cause of lost profit in our business."*

In one of your first meetings (as soon as possible after the job begins), establish a *clear* hierarchy of information flow. Everyone should understand how questions and answers, construction problems, directions, and general information will be channeled and addressed. Are written change requests going to be initiated *only* by the architect? How are plan and specification clarifications or errors to be addressed? This line of communication is important to prevent changes that spring up during the course of construction from being disputed in the name of "unauthorized" direction. Everyone's role must be clearly defined.

At every meeting, update progress schedules and thoroughly address any problems or concerns. If a remedy isn't possible right then and there, establish a firm *next step* before adjourning. Never assume that a problem will simply work itself out, because that seldom happens. When discussing schedules, speak in concise, definitive terms. Never use or accept vague responses such as *"it'll be done the middle of next week"* or *"it'll be shipping around the 6th."* These are simply too open-ended. You want a commitment, like *"Wednesday at 8 am"* or *"it will arrive on site the morning of the 10th."*

Planning and Scheduling

The word *scheduling* can refer to many different things. First, there's the long-term schedule that covers the entire project. This is the schedule that you're asked to submit at the beginning of the job as part of your contractual obligation to the owner. You're also bound to the completion date shown on that schedule. You can create a long-term project schedule using a scheduling software program. I use *PrimaVera*© software for that purpose. Although excellent for presentation to the owner, these long-term schedules are of little use for day-to-day scheduling and planning for materials, equipment, and labor. For daily operations, you should use *short-term* or *short-interval* scheduling.

Short-term scheduling is simply taking a small, fixed period of time, such as the upcoming two-week period, and creating a micro-schedule using more detail and control elements. This short-term schedule should include definitive items such as the actual labor requirements, material delivery dates,

and the scheduling of critical equipment needs (like pumps or conveyors) for the work to be completed during that period. Short-term scheduling helps you avoid those *"early-morning-stumbling-around-locating-tools-and-people"* circus acts that often plague jobs. Eliminating this type of delay is critical in maintaining profit in the field.

By combining both a short-term and a long-term schedule, which still tracks long-range items such as large material deliveries and inspection benchmarks, you'll be able to manage the project more productively. It also helps to make the mammoth undertaking of a commercial construction project a little less overwhelming and easier for everyone to grasp.

There's also a lesser form of scheduling that I simply like to refer to as *daily planning*. This is the everyday common-sense (yet often overlooked) job site planning, such as scheduling your crews so they're not working in the same place at the same time. When workers stumble over one another or are constantly in each other's path, productivity suffers and tempers begin to flare.

And last, there's the placement of material on the job site. I know this one sounds elementary, but I can't tell you how many times I've walked onto job sites where most of the carpentry is being done in one area and the lumber is stacked 85 feet away. And this happens in all trades, not just carpentry. You might argue that a couple of trips back and forth won't do much damage to the budget. But remember, we're talking about the time required for many workers to make many trips. This adds up to a substantial amount of wasted time, and real dollars, by the end of a job. But it's not just the trip itself that does the squandering. There's another predictable result of these treks across the job site. The worker will inevitably stop and chat, or otherwise get distracted or distract someone else, while making his way back and forth — further reducing his and other's productivity.

Leadership

Weak, lax, or unclear leadership is a sure-fire profit-killer. Are we talking about you? The more clearly you define your project goals — *and* communicate those goals to your workers in the field —

the easier it'll be to achieve those goals. However, leadership is more than just preaching that *"productivity equals profit"* at a job project meeting, then adjourning the meeting. Anyone who's read a management book can spout that drivel without even knowing what it means. You need to lead by example. Keep in mind that leadership is contagious. If you are clear and focused about the project goals, your project manager, job superintendent, and even the people in the field will be buoyed by your decisiveness and will follow suit. If every member of the building team is confident and clear about their goals, targets will be met and schedules will fall automatically into line.

Worker Skill and Training

Worker skill is kind of a tricky subject. I said earlier that all crews take about the same amount of time to perform jobs. While that's generally true, there are workers who're consistently more skillful, attentive, and productive than their co-workers. From a productivity standpoint, that means there really is a profit difference between having your "A" crew frame a building versus having your less-experienced "B" team (with workers hired through temp agencies or the local hall) doing the work.

Understanding that these differences exist should be a help in scheduling and assigning work. If a work element is crucial to the continuity of the schedule, you'll probably want to put your best workers on it. However, if the task can be completed at a rate that doesn't put other work or trades at risk, that's generally a good place to assign the "B" team. I'm not saying that all "B" workers are bad news. Many times lower productivity is less about the worker's actual competence as a craftsman than his lack of familiarity with the company or the project. There's a learning curve that comes with any new job, both for you and your tradesmen. After all, it takes time for you to learn about a new worker and how best to use his skills so he's a cohesive part of the team. Having the ability to assess a worker's best attributes is a skill that can in itself create a more productive work atmosphere.

Job-Cost and Field Reporting

Written documentation tracking the progress on the job site is essential. You should keep daily and weekly job logs, minutes of all your job progress meeting (see Figure 11-1), transmittal letters, owner/architect correspondence, and any other written correspondence you may have that concerns the project. I know it's often not easy for your job site superintendents to find the time to stop and write everything down, but it must be done. Good documentation is a habit born of strict personal discipline and a clear policy enforced by the general contractor and company owner.

A thorough written field record will help you handle disputed change orders or claims that pop up once the job is over. It also effectively bolsters overall job organization. An odd side-benefit of keeping careful records is that you'll actually encounter fewer day-to-day disputes. This is because some people (there are one or two on every job) like to throw every change request they can think of at the superintendent. They'll soon learn that you back up each request with a written record. That'll make them less likely to approach with frivolous or petty changes — especially if they'll have to pay for them. Your field log book can be a simple journal or you may use a form similar to the one in Figure 11-2.

Construction Delays

Ask any commercial contractor, *"What's the most frustrating part of your job?"* Nine times out of ten, the response will be: *"Dealing with construction delays."* Interruptions in the project schedule are deadly. They not only mean extra attention and extra labor for the contractor to get everyone back on track, but it means *lost time and money*. Spending money isn't a bad thing — if you'd *planned* on spending it. Unfortunately, most of the time, the cost to correct a delay isn't accounted for in the estimating process, so these costs inevitably reduce your profit.

ABC Contractors

1234 Hammertacker Way, Lienwaiver, WI 53555 • Office (608) 555-1234, Fax (608) 555-1235

Project: Addition & Remodeling, Evangelical Lutheran Church, Janesville, WI

Minutes from March 28th Meeting

Those Attending:

Jim Olson — Evangelical Church
Bill Bannon — Evangelical Church
Dan Wilderhofen — ABC Contractors
Steve Saucerman — ABC Contractors
Bob Walls — Creative Architects, Inc.
Matt Keiner — Keiner Engineering
Bob Walker — ZAP Electric
Ben Sommers — Glugg Plumbing
Ken Wilson — Nicenwarm Heating & Cooling

General Notes

Please refer to ABC Contractors general rules of operation and safety procedures package. If you need a copy, please contact our office at (608) 555-1111.

Meeting minutes contain both old and new business discussed at the prior meeting(s) and include items and questions that occur up to the new meeting. As pertinent items are addressed and resolved, they are removed from the minutes. Pending and/or unresolved items remain.

Business (By CSI Div)

1. Next progress meeting is April 5th @ 9 am.

2. All members are reminded to get insurance certification to ABC as soon as possible and absolutely before their crews are on site.

3. There is an important addenda to this job that includes all of the value-engineering (cost-saving alternates that were taken to meet budget) items for the work. Make sure you receive a copy of these changes, as they affect most scopes of work. Copies will be issued with purchase orders.

4. The church's tax exempt number is E9999-9999.

5. The job foreman is Dan Wilderhofen. His pager number is 608-777-7777. The job trailer number is 608-555-1234.

Schedule (Summary)

Building excavation is complete and footings have been poured. There has been a delay while structural steel column design was approved and the actual columns (required in order to continue with the concrete foundation walls) were produced and delivered. Those columns have now been installed, and the concrete man is scheduled to come back on Monday (today) to continue foundation walls. The foundation walls should take about another week, at which time backfilling and interior underground work will commence.

Figure 11-1
Sample meeting notes

General

1. Discuss upcoming immediate schedule.
2. Status of plumbing, HVAC, and electrical permits.
3. Kevin Canolli has left Creative Architects and Bob Wellwisher will be taking over duties concerning the church.
4. Discuss new entrance at Rockton Road.

Business (By CSI Div)

02

1. Parker Excavating will come back on site as required for backfill.

03

1. Discuss concrete status and upcoming concrete schedule.

04

1. Brick has been ordered and the $350/m was maintained. Scheduled to be delivered on site April 1.

05

1. *See structural steel schedule notes.
2. Upcoming steel delivery status.

06

07

1. Shingle selection — Elk Prestique Hickory.
2. Sheet metal flashing/coping color?

08

1. Wood windows have arrived and are available for installation.
2. Hollow-metal & finish hardware submittals were sent to Creative for approval 3/18/00.
3. Aluminum framing/glass & glazing submittal status.

09

1. Gypsum board & acoustical ceiling submittals were sent to Creative for approval on 2/28/00.

10

1. The church will be buying and installing their own toilet accessories, and credit will be forwarded by ABC.
2. Toilet partition submittals were sent to Creative for approval on 3/1/00.

15.4

1. Plumbing submittal status.

15.6

1. HVAC equipment submittals were sent to Creative for approval on 2/26/00.
2. Remaining HVAC shop drawings/submittals were sent to Creative for approval 3/11/00 (Nicenwarm Brothers).

16

1. Electrical panel shop drawings were submitted to Creative for approval 3/9/00. *Submittals were reviewed by Creative and returned for production.
2. Fire alarm submittals were submitted to Creative for approval 3/14/00. *Submittals were reviewed by Creative and returned for production.
3. Lighting fixture submittals were submitted to Creative for approval 3/18/00.
4. ZAP Electric is reviewing the cost for the future conduit for sound system.
5. Electrical site service status.

Figure 11-1 (continued)
Sample meeting notes

ABC Contractors

1234 Hammertacker Way, Lienwaiver, WI 53555 • Office (608) 555-1234, Fax (608) 555-1235

Daily Construction Log

Project name_____ Date: _____

Owner: _____ Project number: _____ Weather: _____

Cost code	Work description/event	Crew/worker	Manhours	Notes

File: DAYCONLG

Figure 11-2
Daily construction log

The Effects of Delay

A project delay is a slippery thief, silently and steadily encroaching on your company's bottom line. Sometimes it's easy to spot — if you have a busted hose on a backhoe — but other times, the delay isn't so obvious. You may lose an hour or two waiting for a generator to be delivered, or you may have a worker who chronically shows up late and delays the work by 15 or 30 minutes every day. Delays like this affect a job in many different ways. The direct cost of wasted manhours is pretty apparent, but what about the negative impact on morale and productivity? You should consider that as well, because it's quite real.

For example, suppose your superintendent is overseeing a multimillion dollar project and he has the enviable day-to-day responsibility of coordinating 16 to 20 skilled trades (all from different unions), scores of workers, and perhaps hundreds of material deliveries for your project. Now, let's say that each skilled tradesman has his own agenda, attitude and opinions (which he's willing to share freely) regarding virtually every item that negatively affects his work. During the course of a construction project, the interaction between all these parties can get volatile. On any given day, a typical exchange between, say, a journeyman electrician and the job site superintendent might go something like this:

Electrician to super: *"Hey, you! There's a hundred g*@-d*#* (insert the name of another trade here) guys in my way and there's no g*@-d*#* way I can get any of my g*@-d*#* pipe in the g*@-d*#* ceiling until . . ."*

Superintendent (to himself): *"Isn't it time to go home yet?!!!"*

Sound familiar? Your superintendent has his hands full here keeping crews moving and out of each other's way. And if he can't keep them out of each other's way because someone is running a little late here or materials are a little late there, then everyone gets hot under the collar. That slows them down until things are running right again. And this scheduling is only a fraction of the super's duties.

Keep in mind that on top of juggling crews and material deliveries, the owner, architect and even you are making almost daily changes to the super's scope of work, Many of these changes have to be priced, negotiated, fought over, and then approved before proceeding. These are the norm in commercial building, so you can imagine what happens when even a minor unexpected interruption is introduced into the project. You have a delay!

Types of Project Delays

Construction delays create a profit-depleting environment on the job. The most important step in correcting the problem is to recognize that a delay is occurring, or about to occur, *early enough to do something about it.* So let's take a look at some common construction delays and discuss a few methods for handling the inevitable.

Equipment Delay

The only good thing about an equipment delay is that you can recognize it quickly. When a generator doesn't show up, at least you know right away there's a problem because your men aren't making any noise (or they may march into your trailer and make a whole bunch of noise!). Either way, you'll find out there's a problem. The same goes for heavy equipment. At least there's something tangible that you can record and look back on later so you know how to recoup the cost, if need be. When the delay occurs, the job foreman or superintendent should write down any lost manhours or other costs that resulted. You can use forms like the daily equipment log (Figure 11-3) and the weekly equipment summary (Figure 11-4) to help you track your daily equipment needs and plan for the upcoming work.

Major Material Delays

Punctual purchase orders, quick turnaround of shop drawings and submittals, and continuous communication are the major tools for avoiding critical material delays. But, even with the best prevention, there'll be a few product delays that sneak into the mix. Even if these delays seem minor at the time, they can come back to haunt you later on.

ABC Contractors

1234 Hammertacker Way, Lienwaiver, WI 53555 • Office (608) 555-1234, Fax (608) 555-1235

Daily Equipment Log

Project name: _____

Owner: _____   Project number: _____   Weather: _____   Date: _____

Code	Equipment/tool	Own/rent	Work description/task	Hours	Rate	Total cost	Notes

File: DAYEQLOG

Figure 11-3
Daily equipment log

ABC Contractors

1234 Hammertacker Way, Lienwaiver, WI 53555 • Office (608) 555-1234, Fax (608) 555-1235

Weekly Equipment Summary

Project name: _____

Owner: _____

Project number: _____

Weather: _____

Date: _____

Cost code	Equipment/tool	Own/rent	Enter total hours of use for each day below						Total hours (week)	Rate per hour	If weekly rate, enter below Total cost (week)
			M	T	W	Th	F	S/Su			

File: WKEQLOG

Figure 11-4

Weekly equipment summary

Here's a typical example. The wood doors you expected today didn't show up. After several frantic phone calls, you find out they're not coming until *next* Friday — another whole week! And the schedule was going so well. Now your carpenter, who was supposed to be installing doors number 27 and 28 today, will have to knock down, clean up, and leave for another task. He'll have to return next week, set up again and finish the rest of the doors then. That'll cost at least a couple of extra manhours.

But it's not just the carpenter whose work is affected. The painter will have return to stain and varnish the doors another time. Will the painter be happy about this? You know he won't. Most subcontractors would, justifiably, demand additional mobilization money for this extra trip. He probably deserves to be reimbursed, but don't think for a second that the door supplier is going to kick in to cover that additional cost. I've tried that, and it doesn't work. That money is coming out of *your* pocket!

Major material delays can come in many shapes and sizes. Here's another one of my favorite scenarios — one that really happened to me. Our carpet supplier kept our five-month-old purchase order on his desk because he didn't want to order the product too early. He wanted to avoid having to store the carpet in his warehouse for any longer than absolutely necessary. He calls this *"just-in-time"* delivery. I call it *"just-don't-call-me-any-more"* delivery. Why? Because, this is how situations like this always seem to play out:

1. After waiting until "just the right time" to place the order, the carpet supplier calls the factory and finds that the carpet is (surprise, surprise!) on back order.

2. The carpet supplier calls to inform me about the delay and swears it's not his fault.

3. The owner of the project begins to call me at home on evenings and weekends. The terms penalty and liquidated damages begin to find their way into our conversations.

4. I begin to regret not going into the less-stressful occupation of bomb-squad front-man, and swear to myself that I'll never use that carpet supplier again!

Daily Material Delays

Small delays in the arrival of materials are less likely to grab your attention, but very costly just the same. For example, let's say the treated lumber plate material that should have arrived at 7:00 this morning didn't make it in until 7:40. This small delay, though frustrating, may not be worth taking action against the supplier to try to recoup the cost of your lost manhours. But it is worth noting. If it only happens once, you can probably let it go with just a mention to the supplier — that's a reasonable strategy. You get many deliveries over the course of a large job, and many small delays will add up to a lot of lost manhours.

What would be the cost to your company if, over the course of an eight-month-long project, this supplier makes 100 deliveries and half of them are 30 minutes late?

50 deliveries × 0.5 manhours waiting =
25 total manhours

25 manhours × 4 (number of workers) =
100 total manhours

100 manhours × $34.14 (hourly rate) = $3,414.00

When you look at it like that, it's a lot of money! Of course, I'm generalizing a bit. Most of the time these men won't just be standing around waiting for the truck to arrive, they'll still be somewhat productive. Well, maybe. But this illustrates how modest material delays, when spread out over the course of a project, can add up to *real money* by the end of the job. It's always best to have the material delivered at least the night before it's needed on the job. Early morning deliveries have a bad habit of going astray. Figure 11-5 shows a sample daily material log. You can use this or a similar form to plan your daily material needs.

Tools and Equipment Breaking Down

Any seasoned contractor understands the value of consistently performing preventative maintenance on his equipment, whether it's a belt sander or a backhoe. Tools and equipment are a worker's livelihood. And most will agree that cheap equipment is rarely a bargain. Pay for the good stuff and take good

ABC Contractors

1234 Hammertacker Way, Lienwaiver, WI 53555 • Office (608) 555-1234, Fax (608) 555-1235

Daily Material Log

Project name: _____

Owner: _____   Project number: _____   Weather: _____   Date: _____

Code	Product description	New/used ("N" or "U")	Vendor (if applicable)	Work description/task	Delivery time to site	Received/ checked by	Notes

File: DAYMATLG

Figure 11-5
Daily material log

care of it; it's worth it. Burning up motors or replacing switches on inexpensive equipment costs you wasted time and energy. That effort almost always ends up to be more than the $30 you saved going with the bargain brand tool!

Design and Architectural Error

For some reason, I find delays caused by architectural and design errors to be the most frustrating of all delays. Perhaps it's because the architect had months, or even years, to flesh out the details of the design before the project began. Errors on the plans and specifications, along with errors in local code interpretation, are appallingly common in commercial construction today. What's worse is that even in the tight economic market and the aggressively competitive atmosphere that exists in the architectural design business today, the problem shows no sign of getting any better.

For that reason, be prepared to deal with these types of delays. Perhaps the best way to do it is to chronicle all the time lost as a result of design error in a separate daily log book. At the very least, you can use this information as fodder for negotiation later on down the road. It would be nice to charge architectural/engineering firms for lost time, but it's been my experience that a design firm will *never* admit that they're wrong. That's why you'll very seldom be able to recover your losses.

But do keep track of those losses. At the end of the job, you'll have to sit down with the owner and architect to reconcile your final payout (you know, the meeting where they ask you to credit items that *they've* kept track of during the course of construction). Then you can pull out your notes and logs concerning the lost manhours for shoddy design, on-the-fly cost changes, and late or delayed change order billings and say *"Let the negotiations begin!"*

Field Error and Personnel Delay

Of course, contractors aren't perfect either. Errors in the field occur all the time as a result of poor workmanship or faulty judgment, with many manhours lost in the remedy. Since you know that time equals money, you understand that skilled workmanship and quality control aren't simply phrases that make up a good marketing brochure — they're *money in your pocket!* If you don't demand quality from your people, you'll certainly only get what you do ask for.

Another avenue for loss lies in the discipline of the workers themselves. Employees that consistently show up late or seem to "have a hard time getting going" in the morning ultimately dip into the company profit pool. Clear communication and strict employee guidelines are really the only way to plug this leak. Employee motivation is an entity unto itself, requiring sincere, consistent commitment and implementation by the company's owners and managers. It's also paramount in maintaining productivity. We'll examine morale and motivation in detail later on in this book.

Customer Interference

Some customers are *wonderful* to work with, and others aren't. This is a situation that'll never change. Sometimes owners find it necessary to stop out at the site and throw their weight around a little. When this happens, avoid your first impulse, take a deep breath, and then take the time to listen, explain, inform, and even cater to the owner's desire. This may, of course, involve some good acting, and could result in a few minor changes. That won't be fun or productive, but it might be necessary. The owner is probably supposed to run any changes through the architect, but if an item is minor, you'll just be making the customer happy — and a happy customer is good for everyone. However, if the change is something major, you'll need to route it back through the normal architectural change process. We'll look at just how that process works shortly.

Accepting Delays

Project delay is something that'll never go away completely. But if you use clear communication, strict discipline, and consistently enforced rules, policies, and expectations, delays can be minimized.

Using a delay notification like the one shown in Figure 11-6 to put subs, suppliers or workers on notice is an effective means of waking up the

ABC Contractors

1234 Hammertacker Way, Lienwaiver, WI 53555 • Office (608) 555-1234, Fax (608) 555-1235

Delay Notification

_____
(Date)

To: _____   Project Name: _____

_____   Project Number: _____

_____   Owner: _____

Attn: _____   Contract: _____

Notification number 1

As of the above date, your company is causing unnecessary delay to the project schedule for the project listed above. It is imperative that you take whatever steps necessary to get back on schedule by any means available (including overtime, increasing crew size, etc.) at no additional cost to the owner of the project or risk being in violation of your contract and/or having back-charges levied against your contract amount. Please respond to our office within _____ day(s) describing what action will be taken to remedy the situation. Thank you.

Notes: _____

_____

_____

_____

_____

_____

_____

_____

Signed _____

File: DLAYNOTE

Figure 11-6
Delay notification

offending parties. By reinforcing the idea that any delay translates directly into dollars lost, every delay becomes accountable in the eyes of those who perform the actual work. Then they're more likely to share in the obligation to target delays and implement remedies as soon as they become aware of them.

The Change Order Challenge

All jobs are going to have changes, alterations, and/or clarifications during the course of construction. It's quite easy, in the heat of the work, to let minor additions to the job go uncompensated. But, like those lost manhours due to material delays, these small individual changes can add up to very large amounts of money at the end of a lengthy project. That's why the proper administration of change orders is absolutely vital to your commercial construction success.

A Sad (But All-Too-Common) Tale

"Looking back over the job, everything went pretty well. I was happy with the work. It was a fast-paced and harried schedule, as usual, but no major problems. There were a few changes by the owner and architect during the work, and a few more that my guys flushed out that had to be fixed, but nothing that couldn't be absorbed — or so I thought at the time.

"But now, looking over my final job-cost report, I can see that something's wrong. The profit line item on the report is much smaller than I thought it would be. I hoped it was a clerical error, and went over the report again, checking every detail. But the rows of numbers all seem to add up correctly. So, where did the job go wrong? I've strained my brain trying to recall any particular example of waste or any mistake that can explain this profit shortfall, but nothing springs to mind.

"One thing's very clear: the job's over — it's been over for a month — and there's no going back to ask for more money now! Besides, I didn't even know the job was a loser until five minutes ago! How could I? The final billings were still arriving from vendors and subs up until the day before yesterday. And now it's too late. All I can do is clean up the mess, bury the bodies, accept my loss, and search for solutions. Man, how could I let this happen?"

Sound familiar? Probably every contractor who was ever in business has lived through that moment. What do you do? If you're smart, you go back to discover what went wrong. This contractor did. After further investigation, and a chat with his job foreman, he eventually pieced together what happened on that job. All those seemingly minor changes and alterations made during the course of the project weren't quite as minor as he'd thought. In fact, when added up, they turned out to be quite costly. A little labor here, some additional material there, the trim detail that was changed on the fly, and so on. Alone they didn't seem big enough to demand extra compensation, but added together, they came to a total of $4,200!

"Can you believe it? I coughed up $4,200 in extras. My original profit line item was only $10,000, and coupled with some other overages we had, I ended up with just about half of my original profit estimate. I remember that the architect said at the time 'they'd take care of' some of these overages. But now that the job's over, he's suffering from a chronic case of 'selective memory' and can't remember saying anything like that. And, of course, I don't have any written change orders to back me up.

"So I guess I'm stuck. Burned again! From now on, everything's in writing! We'll stop the damn job if we have to! No more 'taking a man at his word' or 'sealing a deal with a handshake' for me! Clearly the idea of the 'gentlemen's agreement' doesn't work! Next time's gonna be different!"

Wouldn't it be nice if we all leaned this lesson the first time? But we're creatures of habit. Some of us learn the first time, but for others it takes a while. This contractor immediately forgot the promises he made himself, and on the very next job, he fell right back into the same old pattern and habits. He'll either have to start using change orders, or he'll go broke. I wonder which will come first?

Changes Are Part of the Business

In commercial construction, it's rare to have a building project completed without any extras. Extras include all the changes that happen *after* the original base contracts for construction are let. Though they occasionally include a credit for an item of work that was deleted or altered, for the most part these changes involve additional charges that should be paid by the owner. Items such as an unanticipated upgrade in the electrical service, the alteration or rerouting of site utilities, or unexpected subterranean encumbrances (like an old building buried on site) are all examples of the type of extras that come up on a job.

If agreed to and accepted by the owner and architect, these changes and their associated costs become modifications to the base contract. Although you sometimes hear them called *contract modifications,* they're generically referred to in the industry as *change orders*. A change order is simply written evidence that an owner has requested the contractor to perform additional labor and/or provide additional materials for improvements which weren't included in the original plans and specifications. Reasons for such changes include:

▌ The owner or architect may simply have a change of heart about one of the details of the project after the work has begun.

▌ The owner may not have clearly understood the architect's original intent/detail as set forth in the original plans or specs.

▌ An alteration may be necessary due to code, covenant, ordinance, or other governing body interpretation.

▌ Errors may be found on the architectural plans/specifications, or details may be poorly defined or not defined at all.

▌ Unanticipated, unforeseen, or unknown site problems or circumstances may impede the progress of the project.

There are other reasons for legitimate changes as well. The key factor in these changes is that they could *not* have been anticipated by any of the parties, especially the contractor, before the original plans and contracts were completed. The commercial contractor who simply "blows his bid" by missing something in the plans and specifications that was clearly defined as part of his scope of work usually can't ask for additional compensation.

The Change Order

Assuming you haven't missed something, underbid the contract, or agreed to some very strict language in the specifications or your contract regarding changes, you're within your rights to request additional compensation should a change become necessary. Of course, you're a nice guy and you want to get along with the everyone, owners and design professionals alike. For some contractors, the simple act of *asking* for additional compensation is very difficult. They don't want to "rock the boat." But situations like those listed above aren't your fault, and you shouldn't have to pay for them.

So you have to ask to be paid for that change order. The success of your company depends on your ability to follow through and collect what's owed to you. Business is business — and like it or not, we're in a field that's grown larger and more structured every day. The small-town relationships of old are now just a faded, sentimental vision. Business people can't just shake hands on a deal, practice the golden rule, and say we'll work out the details later. It doesn't work that way anymore — even if the owner or architect may want you to think otherwise. And, that's probably good, because all sides in a deal, the contractor, architect, and owner, take a business risk when they enter into a large construction contract. They're all susceptible to large losses should unanticipated changes spiral out of control.

That's why the change order is so important. It can mean the difference between profit and loss on your job. The change order is a critical business instrument that demands the same level of attention to detail as the original contract. It is, in fact, a modification of the original contract. I can't understate its importance, especially if the parties have a disagreement some time down the road. Your change order process should closely follow these steps:

1. The request goes to the architectural/engineering firm who works up a formal change order request for pricing.

2. The change order request is given to the general (prime) contractor for pricing.

3. The new price is submitted to the architect who reviews it with the owner's representative and approves or disapproves the change.

4. If the change is approved, it goes back to the general contractor for processing. If it's not approved, the item is generally value engineered, recosted and sent back to the architect for further consideration or final rejection.

5. The general contractor forwards the approved change request order to all subcontractors and suppliers who are affected by the change.

6. All contracts are adjusted accordingly and the work proceeds.

It boils down to one basic rule: change orders *must* be in writing and *must* be signed by the owner, or the owner's representative (possibly the architect).

Figure 11-7 is an example of a change order request which has been priced and is ready to be signed by the owner (or his representative). Then it goes back to the general contractor for his signature and on to the electrical subcontractor for his signature. Copies of the signed document should go to each party involved in the change, including the architect or owner's representative, the general contractor and the subcontractor. Notice that this change order allows you to keep a running total of the change order costs.

The change order log shown in Figure 11-8 is another form that will help you stay on top of change orders. A change order log lets you track all the change orders as they occur throughout the course of the job, their cost and the number of days added to the schedule as the result of each change.

Mastering the Challenge

I've been in your shoes, and I know that we don't work in Wonderland. Projects vary and so do the personality of owners and architects. Some jobs genuinely include a spirit of cooperation and partnering, and that makes the change order process relatively simple and fair to all. But most competitive bid projects don't work that way, so the relationships from beginning to end are adversarial. This type of atmosphere rarely fosters a cooperative mood when it comes to change order negotiations. In a hostile environment it can be quite difficult, if not impossible, to get anyone to sign anything — especially if it involves taking more money out of their pocket.

You may have to persevere, hold your ground, and demand what's rightfully yours. There's a certain skill involved in doing this, and it includes the following:

▌ The courage and conviction to demand compensation for yourself when legitimate changes do arise.

▌ The ability to price, negotiate, and execute these changes *quickly* and *efficiently* during the course of construction.

▌ The ability to enact the changes with as little disruption as possible to the project schedule and continuity.

▌ The tenacity to follow up at the end of the job to insure that all change orders are paid.

This skill must be passed on to your people in the field as well. Part of mastering the change order process lies in the proper instruction and motivation of your job foremen, superintendents, and on-site project managers. Take the time to periodically remind them of the importance of change orders, and the need for consistent enforcement.

Change Orders and Your Staff

All of your decision makers in the office and the field must be fully aware that changes in the field always require full and proper documentation. Verbal changes are never acceptable. Make sure that any field personnel you authorized to make changes always have the necessary forms and documents available to use when they need them. It would be a shame to lose profit simply because the proper form wasn't handy.

ABC Contractors

1234 Hammertacker Way, Lienwaiver, WI 53555 • Office (608) 555-1234, Fax (608) 555-1235

Change Authorization/Contract Modification

Date: September 30, 2000

Project: City Hall Renovation   **To Contractor:** Livewire Electric
              123 Main Street   987 Sullivan Drive
              Hometown, Wisconsin   Hometown, Wisconsin

Contract for: Electrical work   **Change order #:**   1

To subcontractor/supplier:

Your Contract is modified as follows:

9/2/00	Upgrade sub-panel "C" to 400 amp (per your 9/15/00 proposal)	Add	$ 600.00
9/15/00	Phone/data/TV cab, pow/phone to elev, ext outlets, motor watchman	Add	$16,650.00
9/28/00	Various power changes per architect (*see attached letter)	Add	$ 2,002.00
		Total change	$19,252.00

The original contract sum was:	$ 391,500
Net change by previously authorized change orders: _____	$ -0-
The contract sum prior to this change order was:	$ 391,500
The contract sum will be (increased) (decreased) by this change order in the amount of:	$ 19,252
The new contract sum including this change order will be:	$ 410,752

_____   _____   _____
Subcontractor/supplier   General Contractor   Owner's Representative

_____   _____   _____
        Address   Address   Address

_____   _____   _____
Signed by   Date   Signed by   Date   Signed by   Date

If you have any questions regarding this change order, please contact our office.

File: CHNGORDR

Figure 11-7

Change order

ABC Contractors

1234 Hammertacker Way, Lienwaiver, WI 53555 • Office (608) 555-1234, Fax (608) 555-1235

Change Order Log

As of _____

Project name: _____

Owner: _____   Project number: _____

Cost code	Description	Date submitted for approval	Status: A (approved) R (rejected) RS (resubmit)	If approved, enter date of approval	Amount of accepted change ($)	No. days +/- to project schedule	Accumulative total (accepted changes to contract)

File: CHNGLOG

Figure 11-8
Change order log

Try to create a change order program that involves as little time as possible for your people in the field. You want your project managers and superintendents performing the actual construction work — not spending all their time pricing and negotiating change orders. Keep the change documents as simple as possible.

Have your office staff track change orders individually by job, classifying each with some sort of cost code and item type. Choose whatever works with your accounting system. Break out your actual costs into labor, material, subcontract and equipment. Make sure they get done as they come in. Don't let them pile up until the job is completed, or they may snowball on you.

Pricing Change Orders

Pricing the changes punctually and getting those prices to the decision makers is urgent. Most of the time, changes are relatively small and you can price them the same day the request is made. However, quick pricing and turnaround on change requests can be tricky when you're working with an architect on a large commercial project.

In these cases, when a change comes up, you'll often receive a document called *Proposal for Contract Modification*, or something similar. It describes the proposed change and usually requests that you submit a lump-sum cost for this new work. Then you've got to work up your price (which often involves soliciting further pricing from your subs and suppliers), send it back to the architect, who then goes to the owner for approval. Meanwhile, you sit back and await instructions on whether to proceed with the change.

Unfortunately, this process can sometimes take days or even weeks to complete. If the schedule is critical and you can't realistically put together a price in a short period of time, sometimes it's best for all parties to incorporate an unpriced version of a change order. This is known as a *field authorization for change*, and will look something like the one in Figure 11-9. It basically gives the contractor the go-ahead to proceed without interrupting the schedule, and implies an understanding by the owner, architect, and contractor that they'll get together and negotiate

a mutually fair cost settlement later. The decision to go with an unpriced change order will depend on the situation and individuals you're working with. Just remember, this type of agreement leaves the situation a bit open-ended — but that may be the only way to proceed. No one wants to halt the schedule.

There's another alternative you can use when there isn't time to price the changes as fast as they're occurring. Explain that you'll bill the changes using time and materials or hourly fees. Then be forthright and fair when it comes time to bill. This sort of honesty and fair play will often impress an owner, and may possibly win you a repeat customer.

Field Workers and Changes

Encourage your workers to not create changes themselves. This can easily happen when they look at a job and think *"there's a better way to build it."* I see this happen all the time. There are lots of opinions out there, and in construction, there are often many different ways to do something. A worker may have a perfectly acceptable idea about how to do it better, but if the original detail will result in a suitable outcome, tell him to leave it alone! You don't get in trouble for following the plans as often as you do for deviating from them.

Petty Change Orders

Never, ever initiate petty change orders. Don't try to collect for little things like extra local trips, cleaning up someone's mess so you can access your work, or other minor housekeeping. It's just not done. These items fall under the category of the cost of doing business. Be careful not to cross that line between legitimate billable extras and everyday business overhead expenses. Petty charges will almost always end up hurting your chances for future work, and they're not going to pay you for them anyway! Remember, in commercial construction, you're dealing with seasoned business people. They're generally not going to roll over and let you have your way. They can spot a frivolous request a mile off. All you'll succeed in doing with petty charges is alienating the customer and making it more difficult for yourself when a legitimate change request comes up.

ABC Contractors

1234 Hammertacker Way, Lienwaiver, WI 53555 • Office (608) 555-1234, Fax (608) 555-1235

Field Authorization for Change

Project Name: _____

Project Number: _____

Owner: _____

Description of change:

List approximate +/- manhours and material for change

(Note: This information is for reference only and given in good faith in the interest of expediting the schedule. Work will proceed on a time and material basis with final cost being determined upon completion of change.)

Submitted by (contractor representative):   Accepted by (owner representative):

_____   _____
                          Date   Date

File: FLDCHNG

Figure 11-9

Field authorization for change

Markup on Change Orders

On this same theme, I also don't agree with the philosophy that change orders are a good way to pump up your overall profits by attaching inordinately large markups to them. Charging 50, 60, or 100 percent markups on changes and then forcing them down the throat of the owner in the heat of the moment ("We *have* to stay on schedule!") isn't ethical. The customer will eventually figure out he's been nailed, and you'll lose his business! On the other hand, if you mark up changes along the same line as your original percentages, the owner will take notice. The only time a higher markup is justified is when the change involves an excessive amount of time to cost and implement and requires considerable changes in scheduling. Even then, the markup should be fair, not excessive.

Conclusion

You may have noticed that there are a number of items we *didn't* discuss in this chapter that you may have been expecting. We skipped over the cost-cutting methods and money-saving advice associated with things like vehicles and fuel, small tools and blades, or trendy hints about how you can save $6.00 a month on your long-distance bill by switching to Sprint. There's a reason I didn't discuss these items. You see, for years, I've stood by and watched in amazement as well-intentioned (but misguided) construction company owners and managers berated and chastised field personnel for things like using too much gas, burning up too many blades, making too many phone calls, and asking too many questions. And all the while they were missing the far larger business expenses that we just covered in this chapter — like the cost of lost manhours when a worker constantly comes in late.

These managers are either lacking in the ability or the ambition to face the tougher, harder-to-grasp problems like those we've tackled. Or they simply take the easy way out and focus on the trivial and easily-manageable details of their business. The problem is that this nit-picking strategy almost always results in animosity between the company owners and/or managers and the employees. The workers view this behavior as a petty power trip. After all, why should the employees scrimp to save gas or blades when the owner spends hundreds of company dollars on executive lunches, catered golf outings, and other important "networking" or "marketing" events? It doesn't make sense to them!

Overall productivity should always be your goal. Minor infractions are just that, minor. A good work flow and continuity in schedule will earn you many times the dollars that those few extra phone calls cost. If you create a sound business foundation and a work structure that's based on employee involvement and good communication, the small stuff will take care of itself. By creating a productive working atmosphere and implementing some of the strategies we've discussed in this chapter, you'll be well on your way to achieving the efficiency required to generate profit on the job site — and making real strides toward improving your company's bottom line.

Chapter 12

Employee Turnover and Morale

"Rich is a terrific employee. He's been with you for awhile and he's always been reliable and responsible. He gets to work on time, doesn't question orders (at least, not to your face), and lately, he's begun to show some real leadership skills. Even better, the other guys on the crew are beginning to look to him as the leader. You know these kind of guys don't come along every day. And, you also know that lately you've been too swamped with the business end of the company to pay proper attention to your crews. You need a field leader.

"So you weigh the situation: he's got potential, and your work load sure isn't getting any lighter; on the other hand, he's a little rough around the edges, especially business-wise. But the good clearly outweighs the bad. You decide to offer him more of a leadership role in the company. You call him in, let him know your thoughts, and you give him a modest raise and the use of a company truck. You *believe* he leaves your office feeling genuinely enthusiastic and committed to his new role.

"Two weeks later — he quits!

"In your final talk, he relays problems and complaints that you'd never heard from him before — negative things about his coworkers, the company, and you. You take it all in, surprised and confused. You really never knew that there were so many things that bothered him about your firm.

"When he finishes speaking, you just sit there — stunned. You've been blindsided by his totally unanticipated frustration. Somehow, you manage to spit out that you were really unaware that there were so many things bothering him, and you ask why he didn't come to you sooner. You suggest that you could work together to remedy his concerns.

"But it's too late. He's already accepted a position with your competitor across town; and you can tell by the tone of his voice that he's not going to change his mind.

"After he leaves you sit and reflect on what has happened. It becomes clear to you that raises and company trucks may not be enough to keep good people — not the kind of people that your company needs to grow. You have to offer more — but what?"

Investing in Your Employees

As the owner or manager of a commercial contracting company, you take a risk whenever you hire a new employee for your firm. You're betting that the person you're about to select will generate enough profit for your company to justify your investment in him. And the risk doesn't end after you hire and train him, because your investment continues for as long as that person is working for you. You invest in your employees in the form of continuing training, pay raises, bonuses, benefits, and other incentives.

It's this prospect of a long-term investment in another person that often creates a perplexing problem for the commercial contractor/owner. A successful commercial contracting company needs skilled and competent workers to represent it — there's no doubt about that. And to keep these workers competitive and sharp, they have to have their skills periodically updated or enhanced through seminars, schooling, workshops, or other types of organized training. That's the problem.

Once the employee is trained and has acquired greater skill (and greater value in the job market), there's the very real possibility that he'll go elsewhere in search of what seem like better opportunities. It's the old "grass is greener" syndrome, and the ultimate paradox for the business owner. How much training can you afford to give away? When you combine this lack of loyalty with the fact that there's a growing demand for good workers in the commercial construction industry, it doesn't look good for the commercial contractor/owner who's trying to put together an experienced, competent work crew. And, unfortunately, this trend doesn't appear to be improving.

Damned if You Do ....

So what's the answer? Should you stop training your employees? Of course not! If you don't train your people, you might as well give up. And I don't think people like you, who own and operate their own contracting firm, are likely to give up easily.

So, let's tackle the problem. First, it's time for a little investigation. Put yourself in the position of the owner in the story you just read. As an employer, you have to ask, *"Why did Rich leave?"* Your first reaction might be: *"He left for more money — pure and simple."* However, if you dig a little deeper, you may find that there are other, less obvious, reasons why Rich left. And those reasons have less to do with money and more to do with morale and self-esteem. Over time, Rich's morale had slowly been eroded away, undermining his job motivation and his ability to remain content at work.

"But I gave him a promotion, a raise, and a company truck," you might say. *"Just how much self-esteem does a guy need?"* Well, the answer, unfortunately for you, is a lot more than you probably think. Though offering Rich a promotion in your company seemed a grand gesture to you, it may have been (and here's the hard part to swallow) too little, too late. By the time the offer was made, it obviously wasn't grand enough to keep Rich content, satis-

fied, and loyal to your company. You only *thought* that it was enough. You didn't actually discuss his future or his feeling about the job — you just threw him a bone or two. The result was another employee turnover statistic.

The Real Problem

If you're sincerely going to tackle the problem of turnover, one of the first things you're going to need to do is acknowledge that turnover is a real, profit-depleting business liability. It encroaches on your company's bottom line just as much as a missed line item on an estimate or having to correct faulty carpentry work. The only difference is that turnover is far less tangible, which makes it harder to address and correct. To track your turnover rate, use the tracking form in Figure 12-1.

You've got to recognize that excessive employee turnover is almost always a result of poor worker morale. Attitude and motivation are essential for success. They're associated with virtually every aspect of an employee's ability to perform his work. If you're going to make serious inroads into curtailing turnover, you're going to have to address the root cause of the poor morale. This isn't easy. You'll have to be sincere in your efforts, patient, and consistent in providing a remedy.

Looking Inward

How do you know if you have a problem? Take a look around your company and ask yourself the following questions:

1. Are people leaving the company with greater frequency than before?

2. Has there been an increase in employee complaints or health claims?

3. Have you noticed greater absenteeism and poorer punctuality by workers?

4. Have there been more spontaneous outbursts or disruptions by workers?

5. Has the quality of the work declined?

6. Do your workers appear to care about their work less than they used to?

ABC Contractors
1234 Hammertacker Way, Lienwaiver, WI 53555 • Office (608) 555-1234, Fax (608) 555-1235

Employee Turnover Tracking Report

For the month ending _____

Employee description/department	Average number of employees	Turnover number (month)	Percent of total department
Administrative/managers			
Estimators/project managers			
Office/clerical			
Design/drafting			
Field - supervision/foreman			
Field - cement finishers			
Field - carpenters			
Field - laborers			
Other			
Totals			

Year to date _____ **through** _____

Employee description/department	Average number of employees	Turnover number (month)	Percent of total department
Administrative/managers			
Estimators/project managers			
Office/clerical			
Design/drafting			
Field - supervision/foreman			
Field - cement finishers			
Field - carpenters			
Field - laborers			
Other			
Totals			

File: EMPLOYEE

Figure 12-1
Employee turnover report

If the answers to any or all of these questions alarm you, don't panic — but do act quickly! Poor morale doesn't just go away by itself. In fact, it often works in exactly the opposite way. It festers, grows, and spreads, contaminating everyone in its path. It becomes a fertile breeding ground for employee discontent and turnover.

Think this just a tempest in a teapot? Think "feelings" are for wimps and whiners (after all, you're paying them to swing a hammer, not for them to have warm fuzzy feelings about you). Well, consider these points:

- Employee turnover disrupts continuity in the field. When new hires replace workers who are familiar with a project, the whole job slows down. You've got to repeat old communications and directions. And this lack of efficiency can not only reduce your profit on this job, it can also jeopardize your future work. When the word gets out to prospective clients that your crews are "suspect" (and trust me, the word *does* get out), you'll find that you'll be getting fewer job offers. No one wants to take a chance on an unstable work crew.

- Let's face it — you're probably aren't running a company like Microsoft. You don't control your entire market area. You probably have between two and four major competitors in your region. When you lose an employee, he usually ends up with one of those competitors, taking your company secrets and strategies. Even worse, he'll take your weaknesses and/or problems — information your competitors can easily exploit.

- And it can get worse. Unfortunately, there's often animosity between the departing employee and the company management, so negative rumors ride along with the company secrets. Over time, these rumors get repeated, which makes them seem more valid. Eventually, they find their way to your prospective customers. And this can have a very negative impact on your company's bottom line.

- Customers become comfortable working with particular employees from your company — especially if they're repeat customers or if the work extends over long periods of time. When one of these familiar faces leaves your company, the owner can feel abandoned. No one likes to break in a new contact. This is particularly damaging when the type of work is very technical or specialized, as when you're working with hospitals, health care facilities or manufacturing clients.

- If you have disruptions due to staff turnover, you may tend to be less aggressive in going after certain types of work. Concerns over staffing or reservations about whether your productivity and quality are up to par can greatly influence your decision about how much and what kind of future work you should pursue.

- A revolving office door can have a strong negative effect on those employees who stay. Not only do they have to train and adapt to new people, but they've got to wonder *why* the turnover is taking place. They begin to ask themselves why they're still there when so many others have left. Is there some opportunity out there that they're missing? *Should they start looking around, too?*

- Finally, there's the enormous amount of office time and energy, not to mention advertising dollars, that go into finding new employees. Someone has to write ads, take calls, distribute applications, consider resumes, schedule the interviews, meet with the prospects, make a selection, and then train the new hire. (Oh, the dreaded learning curve!) If you don't believe all these things are costing you money, you're fooling yourself.

Getting a handle on employee attitudes is tricky — and a formidable challenge for even seasoned business owners. Attitude problems can't be as easily addressed as material or equipment problems. If the copier breaks, you fix it; when office supplies run out, you replace them. But employee morale can change daily. Workers who seem content on Monday may be disgruntled on Thursday — and the cancer can spread fast. A young employee hears an older one criticizing management and, searching for acceptance and common ground with his new coworkers, he begins to echo and believe the sentiments he hears. Soon the echo is a rumble, and then a roar.

It doesn't matter if the criticism that started the echo is justified. There's a strange group mentality that comes over workers when they hear these things. *Anything repeated over and over again — no matter how ill-founded or incorrect — is very often accepted as fact by some of the listeners.* This same phenomena explains why many of us own Ronco® Food Dehydrators, Ginzu® Steak Knives, or the complete set of Banjo® Fishing Lures. After hearing the sales pitch over and over, the product begins to sound very good, and you have to have it!

When the negative group mentality takes over, you'll suddenly find that efficiency is suffering. The situation may be about to spiral out of control. But what can you do? You're a contractor — not a psychologist!

Taming the Beast

There *are* things you can do; simple, effective, even economical things. You need to remember, however, that the human psyche is infinite in design and complexity, and what works for one person may not work for another. What we'll discuss here are just the basics. We'll look at a few relatively straightforward methods for getting morale under control — methods that are well within the reach of the average commercial contractor/owner.

Getting the Word Out

I've never been one to call a lot of meetings. I don't like them, as a rule. Meetings have a way of taking off into unplanned territory or disintegrating into idle prattle. But there are times when meetings really do make good sense — and addressing a morale problem is one of those times.

Call a meeting of your employees and let them know that the express purpose of the meeting is to discuss morale and attitude — *and nothing else*. Let them know that you would appreciate open, candid, and honest input regarding the morale problem in the company. Explain that you're trying to understand the situation and that none of their comments will be held against them. And, by all means, honor your word on that subject. Things may be said that'll hurt other people's feelings — or your feelings, for that matter. It's OK. It comes with the territory. Isn't it better to air grievances in a controlled, open forum rather than hearing them from a third party, or worse yet, hearing them on the job site in front of your customer?

Once you get your employees talking, *listen to what they have to say*. Some suggestions will be good, some bad, some petty — and others just plain dumb. Whatever you do, don't belittle anyone's suggestion or make the person feel stupid. What may seem stupid to you now might make a whole lot of sense if you were on the other side of the desk. And don't take anything personally. In the end, you'll probably be surprised by what you hear. Contrary to popular belief, there really are some bright and enlightened people out there. And most of them want the company they work for to succeed. They want to be proud of their company. No one wants to work at a business that's second rate.

Then carefully weigh the information you get at the meeting. Don't think for a moment that I advocate caving in to every demand presented by your employees. You're still in charge and you still make the decisions. But you also have the most to lose in this situation. Remember that most of what you hear may be nothing more than therapeutic venting. But by listening to them, you'll show your people that you take their input seriously. The opinions you get from them will make it easier to come up with a possible remedy.

Discuss the ideas that you feel worthy of pursuing openly and sincerely. For areas that require further action, set specific goals, or at the very least, define the next step you need to take to proceed toward a remedy. Don't take it all on yourself. Make the employees part of the solution by involving them.

Ask them how they'd handle the situation and how they'd proceed. Mix your own ideas in with theirs, and you're on your way to a solution — and you're working as a team. Employee involvement is a key part of the solution. Besides gaining assistance, you'll also be taking your first step towards *empowering* your employees. Yes, I know this sounds flaky, but it's a real tool that can make an enormous difference in your company's success.

Empowerment

I'm a huge believer in empowerment, or more specifically, empowering certain employees with greater responsibilities within the company. Empowerment is simply the act of increasing the level of authority of selected employees. It used to be called *delegating authority*. Today, it's called smart business! When you empower an employee, like your job superintendent or foreman, you give him greater responsibility for his actions and more latitude in making his own decisions. That also means you give him more leeway in making his own mistakes as well.

Empowerment is a strange thing. By relinquishing just one level of your total command, you're telling that employee that you acknowledge his skill, trust in his judgment, and believe in his abilities. This can be a huge motivator for people. Their pride suddenly kicks in, and they try much harder to do everything right because their decisions and actions affect the end product. They become *more responsible*, *more inventive*, and *more attentive* to their work. In addition, they take pride in their role with the company, which makes them more loyal. Their new responsibilities give them a greater sense of importance, and a boost to their self esteem. And you did all this for them in the name of company efficiency!

Why do I say all in the name of efficiency? Because you really can't do it all — even though we all try. There comes a time when even your best effort can't get everything done, and you know you have to ask for help. There are too many decisions to be made on a daily basis. You have to delegate some

> *"Your employees are arguably your greatest asset in achieving your profit goals (or your biggest stumbling block, if their attitudes turn negative)."*

of your authority. I know that this can be quite difficult for some of you contractor/owners out there. It's hard to let go. That's understandable. After all, you probably built your company from the ground up. You naturally have a strong emotional attachment to every aspect of the business, and you've grown used to taking on all challenges by yourself. But face it, how much can one person do?

Your company is growing, and by empowering others, you not only improve their self esteem, you also free up more of your time. That'll actually give you the ability to pursue some areas of your business that you've probably been neglecting, like networking, marketing, or expanding your territory. Maybe you've considered participating in community work (besides being good for the community, volunteering often provides the type of business connections that lead to jobs). In short, you can have more time to do the long-range planning that you never seem to have time for. Those things are important in determining your company's future. And you *must* look ahead if you're going to be successful.

Reward Programs

You can't really talk about improving morale without discussing reward programs. People are people, and everyone likes to be rewarded for a job well done. There's really no greater motivation. And that's how it should be. If an employee's work is above and beyond the call of duty, shouldn't he be rewarded? Of course! Now, I'm sure you've read articles on morale that say there are methods of reward other than money. Of course there are other rewards, but are they as effective as cash? I doubt it! Money is the greatest motivator. Given the choice between praise and $500, I'll take the $500 every time! Maybe when the bank will accept praise in lieu of the next payment, I'll reconsider.

It's not simply greed. If you're an average Joe like me, your absolute first priority in life is to provide for your family and to make sure that their needs are met. And something intangible that makes you feel good just doesn't do that! Praise and other small rewards are wonderful supplements to money, but money is what most people need the most. And from a business perspective, I couldn't think of a better place to spend your money than on your employees. They're your best tools! Your employees are arguably your greatest asset in achieving your profit goals (or your biggest stumbling block, if their attitudes turn negative). Yet, many employers still hold back when it comes to handing out the checks.

Why? That's a tough one to answer. The reasons vary from personality to personality, but here are a couple I've noticed along the way:

▌ I think the single biggest reason that bonuses aren't handed out more frequently is due to the greed or ego of the owner. I'm trying to be candid — not cruel. But, it's true. Lots of owners clearly look out for number one, and have a hard time letting go of their money. This is tough to address. You simply have to be honest with yourself and ask, *"Am I that kind of owner?"* If the answer is yes, I hope you'll take this advise to heart and realize that everyone will prosper if your company runs at optimum level — and you can only achieve that with the help of motivated employees.

▌ The second most common reason involves the notion of "parity" among the employees. It's an incredibly skewed line of reasoning that goes something like this:

"I know you work harder than the rest of the guys, and I know you've developed better skills. I also know that you have a better attitude about your work than the others — but if I gave you a bonus, all the other employees will want bonuses, and I just don't have that kind of money to hand out!"

Huh? Why would any of the other employees even *think* they deserved a bonus? Have they earned it? Where's the motivation for an employee to work hard and do their best if there's no reward? The only thing that this kind of thinking encourages is a slow death-march toward mediocrity. Why would anyone try to rise above the ordinary if there's no reward for it?

Is this the kind of attitude you want to encourage?

The Bonus

For those of you who are still finding it hard to loosen your grip on the bonus money, let's talk about the bonus itself. Over the course of my construction career, I've worked for or witnessed owners who, though not willing to part with *a dime* to provide incentives to deserving employees, had no difficulty spending money on items like these:

▌ $900 for a potted plant for owner's office.

▌ Hundreds, possibly thousands, of dollars for sporting events and theater tickets to entertain "clients."

▌ $9,000 for a glass bowl to be displayed on a shelf in the reception area.

▌ Generous salaries for family members, who have never even been to the office.

▌ Thousands of dollars on fad software programs and computer hardware, most of which never even made it out of the box.

▌ Expensive cars, travel, dinners, exclusive club memberships, and golfing events which were written off as business expenses — and much, much more.

I think you catch my drift. If I see money being spent like this, what makes you think that your employees don't see it as well? There's really no justification for this lopsided distribution of wealth. You have to decide whether or not the people who perform your work are the most important attribute of your company. (Hint: *they are!*) If you believe they are, then the road you're going to take is clear and fair. If you *don't* believe your employees are your greatest asset, then you'll eventually fail.

Don't become another business statistic. Take steps to address the issue of morale before it becomes a problem. Company morale can be a positive, profit-making tool for your business — but only if you want it to!

A Course of Action

There'll always be a handful of "pass-through" employees, like students, who'll only want to work for a short time. And there are others who'll come and go who really shouldn't be in the industry in the first place. Turnover among that group is a blessing for all. But for the majority of workers who are worth keeping, here are a few morale-building techniques you can use in your company to help keep your turnover rate down.

1. Acknowledge that your employees have a life outside the office. Don't make unreasonable demands on their after-hours time. You'll have a few emergencies and pressing problems that require extra time. And there'll be the occasional bid that *has* to get out the door by a certain time. Employees understand these types of demands. But don't be one of those bosses who have chronic "urgency addiction" — who want everything, even those things that can wait, done immediately. Employers like that make their employees feel that they can never sacrifice enough for the company, even if they work evenings and weekends. Sooner or later the employees get tired of that kind of treatment and look elsewhere.

2. Always treat employees as "permanent," even if they're hired on as temporary crew members and you know you'll have to lay them off when the job ends. If an employee is preoccupied with whether or not he'll be let go soon, his work will understandably reflect it. If he thinks there's a chance you might keep him on, he'll perform better and have a more positive attitude toward the job. And, who knows, you might be impressed with his work and want to keep him.

3. You don't always have to pay more to keep good employees. Sometimes the benefits package you offer translates into more money than a higher hourly wage. Health insurance, paid vacations, and a retirement program can go a long way toward keeping people with your company. I know that most employees have come to expect these benefits as a standard part of the employment contract, but it never hurts to remind them that you're giving them more than just a wage. Let them know what you have to pay to keep up their health insurance. Many employees have no idea what insurance costs their employers, or how much they'd have to shell out if they didn't have insurance! Knowing the true value of their wage may give them a new-found respect for the job.

4. Offer occasional time off for personal business (with no strings attached), particularly in slow business times.

5. Commend your employees on a job well done. Some managers seldom or never offer any verbal encouragement. A simple "thank you" can make a world of difference to some people. But use it *in addition to* money, never instead of.

6. I know some people think company-sponsored outings like picnics, holiday parties, or pizza lunch on Friday are a little hokey, but a lot of employees really enjoy occasionally socializing with their fellow workers. These get-togethers are usually appreciated, especially by employees with young families.

7. Occasionally offer the use of a company vehicle for an employee's weekend move or a heavy project around their home. I know there are possible liability, insurance, and IRS ramifications, but it doesn't hurt to help people out once in a while. Make your own decision about this one — only you know what's important to you and your people.

8. Have employees switch jobs from time to time. The benefits work both ways. The employee doesn't grow stagnant and bored from repeating the same work every day, and you'll have an employee with additional skills, offering you more flexibility when scheduling tasks.

9. Recognition programs (certificates, pins, plaques, etc.) are great morale boosters — *but not if they're substitutes for raises or bonuses!* A common mistake by many employers is having only a certificate or pin for their entire incentive program. People aren't stupid. In the absence of a more substantial reward, such as a bonus, promotion, or prize of some sort, a certificate can come off as miserly, condescending and superficial.

10. Provide your employees with an avenue for personal growth and development. Everybody, to some extent, wants to grow and learn. It's natural to want to expand your horizons. Spring for the occasional seminar or pay for a night school course. It won't cost much, and again, everyone benefits. You gain a worker with enhanced skills, and the employee feels as though he's moving ahead in life and isn't just stuck in the same old routine.

11. To give your company a recognizable personality, create a monthly newsletter with information about the company's goals and progress, upcoming seminars, company outings, etc. Invite the employees to contribute ideas, editorials or cartoons. Let them share in the company vision. Ask for ideas on ways to improve your product and service. You may be surprised at the level of suggestions you get. Hold a contest to come up with better ways of doing a job, or to create a company mascot or new logo. Award prizes — anything from a cash bonus to dinner out for two. The newsletter wouldn't have to be elaborate or long. One or two pages would be plenty. You don't need to go to a printer — these days, you can create very professional-looking newsletters with just a word-processing program or a simple desktop publishing software program and a color printer. All you need to provide is the content.

12. When creating a new position in your company, try to promote from within. Don't run out and solicit people from the outside in the hopes of finding a more "perfect" employee for the job. Stop for a moment and take an unbiased look at the people you already have. There's no better way to show an employee that you sincerely appreciate their loyalty, dedication, and efficiency than to offer them a promotion within your company.

13. Know what your turnover rate is. Don't forget that tracking report back in Figure 12-1. You may be surprised at the number of employees that come and go. If you notice a sudden upswing in turnover, especially in one department or area, it may be time to step up your morale program.

14. Perform exit interviews with employees who leave your company. You can use the form in Figure 12-2 as a guide when conducting these talks. This is similar to the morale meetings we discussed earlier — you may not always like what you hear, but it will do everyone good if the employee will discuss problems openly. This is a very good way to discover what employees *really think*. After all, they don't have any reason not to be frank and honest with you if they're leaving anyway.

Conclusion

There are many ways to build morale within your company. Morale-building remedies can be as simple as a company-sponsored picnic, or a pat on the back. What about offering to pay for athletic club memberships as part of your health care package? Or sponsor a company softball team? Ever considered developing your own in-house training sessions? You're truly only limited by your own imagination and your willingness to make a difference.

Build a bond with your employees. Let them know you consider them to be more than just "tools of the trade." By allowing some of your better people

ABC Contractors

1234 Hammertacker Way, Lienwaiver, WI 53555 • Office (608) 555-1234, Fax (608) 555-1235

Employee Exit Interview

Employee name	Employee number	Department

Type of termination	Termination date	Today's date

Parting employee's comments, suggestions, observations:

_____

_____

_____

_____

_____

_____

_____

_____

_____

_____

_____

Company interviewer's comments:

_____

_____

_____

_____

_____

_____

_____

_____

_____

_____

Interviewer _____   Department _____

File: EXIT

Figure 12-2
Exit interview form

a bit more authority and control over their work environment, you'll be relieved of some of your responsibility. And that's good for everyone, right? As you improve your company, you also improve your people and their relationship to your company. Start making positive changes now, instead of waiting until problems begin to crop up. Once they do, it's often too late to slow the momentum, and the damage is done. If you're sincere, patient, and committed to your program, turnover will go down, morale will go up, and productivity, efficiency and profit will follow.

Chapter 13

Project Closeout and Warranty

There's arguably no greater satisfaction for a commercial contractor than completing a construction project — and I mean being *really finished*. Unfortunately, in many commercial construction projects, being *really finished* is open-ended, because everyone defines "finished" differently. So the owner, the architect, and the contractor, looking at the same project, may all have a different idea about when the project is *really finished*. Let's look at an example of an open-ended project.

The Job That Wouldn't Die

"A week after completing all the items on what you thought was the one and only punch list for the project you've been working on, you mail your final bill to the owner by way of the architect. On that very same afternoon, you find in *your* mail, quite unannounced and unanticipated, a *second* punch list from the architect. It contains a new list of work items that are "required to be remedied before final payment can be released." You thumb through the multipage document and take in the contents.

"The new punch list describes numerous nicks, scrapes, and scratches on walls, doors, and floors throughout the project. This damage may very well have been caused by the tenants who moved into the building at the beginning of the week, but how can you prove it? You didn't see it and it wasn't on the original punch list — but it's there now. It's possible that everyone overlooked these items. It's a large project.

"The list goes on to include other larger, more-involved problems that will clearly require a fair amount of time, attention, and money to remedy — money that wasn't covered in your bid! You should have known this would happen — it always does on big jobs. With all the discrepancies you had to interpret on the plans, it was inevitable that one or two would come back to haunt you in the end. Not that you didn't try to clear everything that seemed really big and important with the architectural firm — but you just can't keep a schedule and double-check on every single minor detail! There are just too many. So you decided where the chalkboards should be placed, which accessories should go in which bathroom, what color grout to use with the tile, and how high the coat racks should be mounted, etc. These weren't big decisions, and most of them worked out, but correcting the ones that didn't will cost time and money.

"You weigh the situation in your mind. You genuinely believe the second list is unfair, but don't want to risk alienating the owner. You hope to do more business with him in the future. And for that matter, you don't want to anger the architect, either — he has the ability to bring a lot of bidding opportunities your way. They both probably know this, or they wouldn't try this second punch list. But you decide that doing this work, even though you shouldn't have to, will cost you less in the long run than losing two very good future sources of income. So, you count to ten (to lower your blood pressure) and accept this frustrating turn of events.

"You spend the afternoon in your office making phone calls. First you phone your job foreman, who's now busy with a new project, to let him know that the other job isn't quite done yet. Then you break down the list of problems by trade and call the subs and suppliers who are impacted by these new demands. They're not happy when they hear about a second list. However, after some cajoling and convincing, you get them all to come back. You're still holding the trump card — their money! You set up a schedule to complete the items on the new punch list.

"It takes two weeks to finish the second list. And, as luck would have it, the architect is out of town as you're finishing up, so you have to wait an additional week to do another punch list walk-through. Meanwhile, the final billing you mailed has been sitting on the architect's desk for almost a month, just as most of the subcontractor and supplier invoices that you received at the end of the job are sitting on your desk — unpaid!

"The owner/architect walk-through finally comes off, but two *more* weeks go by before he approves your final pay request and sends the package on to the owner for payment. He wanted to go over everything just one more time — and he's a very busy man, you know. He's withheld a contingency for a "few concerns" that he still has about some of the work. But, you figure, at least you'll have most of your money and you can get this job behind you.

"Murphy's Law being what it is, the board of directors of the owner's company, who must approve all payments, only meets on the second Monday of every month, and your payment request *just missed* this month's meeting. *"I guess it'll just have to wait until next month,"* the owner calmly tells you.

"By now, you're fit to be tied. Your subs and suppliers are calling daily and growing angrier every time they call. They need their money. A few have even threatened a lien against the job. What can you do? You have to pay them off out of your own funds.

"The next month finally rolls around and the board approves your final payment and sends it to their accounting office to process. Success at last — well, sort of. It's another 45 days before the check finally arrives at your office! *"Yes, I know your original agreement said 'payment is due 10 days after final approval.' So sue us!"*

The Joys of Contracting

Frustrating? You bet! But the truly scary part is that this example represents only one of the ways that an otherwise profitable and successful commercial construction job can become derailed in its final stages. There are many closeout requirements for commercial projects that, if left unsatisfied, can potentially plug up the money conduit. And unfortunately, at the end of the job your cash flow can be critical. By the time you get to this point, you have a lot invested in the project. You've expended tremendous amounts of time and effort, not to mention any money you may have extended to cover subcontractors and suppliers who couldn't be put off. It's not unheard of for a contractor to be fiscally in the red at the end of a project — and that's no time to lose control of the revenue flow!

To ensure that the payments keep coming, especially that all-too-crucial last payment, it's critical that you understand and satisfy all your project closeout responsibilities. You must have procedures in place to handle closeout administration, and you must be able to anticipate potential eleventh-hour roadblocks. By preparing yourself to efficiently handle any possible problems that could pop up at the end of a job, you may be able to prevent a delay in your final payout. And more than likely, that's the one that includes your project's entire profit!

Project Closeout Requirements

Let's look at the commercial construction closeout process and see how you can gain control over those elements that can mean the difference between getting your money *now*, or getting it *later*.

The Punch List

The *punch list* is probably the most familiar part of the closeout procedure for most of you reading this book. For those who have never had to deal with this experience, the project *punch list* is a list, usually created by the owner and architect at the end of a job, that indicates any incomplete, or *deficient*, items of work on the project. These items are normally required to be completed and/or remedied by the contractor before his final payout is released. This obviously makes the punch list extremely important, especially to the contractor. It also can be the cause of some heated disputes during the closeout period.

And yet the disputes, when they do arise, aren't centered so much around the fact that there's a punch list, but rather they're about the punch list procedure. Most contractors understand the importance of fulfilling their contractual obligation, which is to deliver a completed, quality project, including all the final details, for the agreed-upon price. However, disagreements tend to spring up over poorly-defined or inconsistent demands made by the architect or owner which carry over into one punch list after another.

No one minds finishing up the details — but this sometimes gets ridiculous. I've had jobs with as many as four or five different punch lists, some containing phrases such as *"it doesn't appear . . ."* or *"it doesn't seem to be what the plans and specs had intended."* But what it really means is it isn't exactly what the architect had in mind, but he can't specifically tell you what that is. He just wants you to redo it over and over until you match some vague vision of what he thinks looks right. Apparently, somewhere along the line, *mind-reading* became a prerequisite for becoming a commercial construction contractor!

No, this isn't a cheap shot. It's a very real problem for commercial contractors — and a problem that has a marked effect on your bottom line. The punch list exercise is simply too discretionary, too individually alterable, and too dependant upon the personalities of those making up the list. It can contain items that range from totally justifiable to absolutely ludicrous. But, even if you *can* prove that a particular item is absurd, you're not out of the woods. Many an architect, unwilling to admit to an error in front of the owner, will simply fall back on the classic *"I know it may not make sense to you, but I'm the architect!"* The implication, of course, is

that you're too stupid for words and you'll never understand, so just do it or you won't get paid.

Unfortunately, that ploy usually works!

The average commercial contractor simply wants a list that he can finish! It needs to be specific, defined, and predictable. To protect yourself and your company, insist that there be only *one* punch list — and hold firm in your stance. If this means the owner and architect need more time to work up their (one and only) list, that's fine, let them have it. Then, once that list is satisfied, have the architect or owner (preferably both) sign off on the work. This eliminates the possibility of them coming back later to have you repair questionable scratches and dings. It's also a good idea to run your own walk-through with your crew a day or so prior to the actual punch list review by the architect and owner. You can compile your own punch list using a form like the one in Figure 13-1. This "dry run" can help you target and anticipate work items that need more attention before the real closing commences.

The one-list policy is something that you'll need to define early on in your negotiations — to avoid any conflict or confusion later. If the architect and/or owner won't commit to it, then you might want to consider adding some money to your bid to cover the almost-guaranteed additional expenses that you'll have coming at you at the end of the project. You may even want to reconsider taking the job at all. Don't allow yourself to be intimidated. There's plenty of work out there for good contractors. Part of being successful in business, and in life, is having the ability to set your own ground rules rather then letting others set them for you! Take control. Let them know that their (one) punch list really is *the conclusion* of the work.

Project Record Drawings and Documents

At the end of most commercial building projects, the contractor is required to submit to the owner a set (or sets) of record drawings that illustrate any deviations to the plans and specifications that may have occurred during the course of construction. These record drawings are referred to as *as-built*

ABC Contractors

1234 Hammertacker Way, Lienwaiver, WI 53555 • Office (608) 555-1234, Fax (608) 555-1235

Project Punch List

Project name: _____ Date: _____

Owner: _____ Project number: _____ Architect: _____

Work to be completed by _____

No.	Cost code	Item description	Corrected by:	Approved by:	Notes
1					
2					
3					
4					
5					
6					
7					
8					
9					
10					
11					
12					
13					
14					
15					
16					
17					
18					
19					
20					
21					
22					
23					
24					
25					
26					
27					

File: PUNCH

Figure 13-1
Punch list

drawings, or *as-builts* for short. They become an important reference document for future occupants of the project. Almost all jobs encounter some changes along the way, some more than others, and these modifications can be of critical importance if there are any changes in or around the buildings in the future. For example, what if the gas line that's shown in the east wall of the kitchen on the original plans is now in the north wall, the wall that the new contractor is about to tear out to expand the kitchen area?

Generally, as-built procedure calls for each trade (plumbers, electricians, carpenters, etc.) to take an original set of plans and specifications and mark in red (or additional colors as appropriate for clarity) any and all changes that occurred during the course of construction related to the scope of their work. The prime contractor is then responsible to gather up the individual trades' as-builts, as well as his own, and compile a master set of record drawings to be turned over to the owner as part of the closeout requirements.

Operations and Maintenance Manuals

Often referred to as O & M manuals, Operations and Maintenance manuals are usually a collection of technical information, in a three-ring binder divided by subject, describing the primary plumbing, mechanical and electrical systems and equipment that have been installed in the project. Heating, air-conditioning and ventilation (HVAC) and electrical control packages are common candidates for O & M manuals. They include critical information related to the installed systems, such as:

▌ emergency instructions and procedures

▌ maintenance agreements

▌ maintenance protocol

▌ parts lists and ordering information for maintenance

▌ wiring diagrams and schematics

▌ cycling and inspection schedules and dates

▌ testing and balancing reports

▌ fixture lamping schedules for lighting

▌ general product data and literature

▌ warranty information and dates

▌ certifications

▌ information relevant to the operation of the equipment

Shop drawings and submittal packages, generated early on in the construction phase, may also be included in the O & M manual. These manuals, like the as-built drawings, remain with the building for use by the future occupants. They're crucial reference materials for those maintaining the facility or building.

Warranties

There are generally two kinds of warranties that you'll be required to provide as part of your closeout procedure: *product warranties* and *labor warranties*. Warranting your labor is usually just a matter of writing a letter that states that your company will guarantee the quality of your workmanship for a specific period of time. You'll also need to obtain similar letters from your subcontractors. Then you and your subs *must honor* those commitments. Though one year after project completion is the common warranty period for labor, you could be required by the original project specification to provide coverage for a shorter, or more often a longer, period of time.

Collecting, assembling, and delivering product warranties can be a bit more involved, but it's still a relatively straightforward process. Product warranty information is often provided along with products or equipment when they're delivered, or you can get them through the product's manufacturer. Warranty information is commonly included in the shop drawing/submittal packages from subs and suppliers. For the general contractor, the product warranty process involves gathering up these individual warranty items and organizing them, usually by CSI division, so that they're available for future owners.

Final Inspection

You'll have inspections taking place throughout the construction process, but at some point, you need to have final inspection documentation. In some places, a number of different inspections are required, often involving different inspectors. Large cities, for example, may have different inspectors for general building, plumbing, HVAC, electrical, fire safety or other divisions. In addition to these, there may be required state inspections, as is often the case with elevator installations. The final inspection often encompasses all of the phases at once, but it can be done in phases. The final inspection documentation goes into the closeout package for the owner.

At the end of a commercial project, the name of the game is occupancy — *owner occupancy* that is. They want it, and you control it! In addition to the regular inspections, you'll likely be required to furnish an *occupancy permit*, or in some cases a *partial occupancy permit* if a project is developed in segmental phases, from the local governing code body. These inspections are usually scheduled alongside the regular local inspections and the documentation is collected and forwarded to the local code body. This assures the owner that you've complied with all of the rules and regulations and that the project is ready to be inhabited. There may also be additional operating permits, certifications, or other releases required before occupancy can take place. Check with your local code and zoning authorities.

Final Cleaning

In most commercial construction specifications, the prime or general contractor is required to perform the final cleaning of the premises as part of his scope of work. This includes the building(s) and associated building site, and may also include adjacent properties if debris from your building spills over onto those lots.

Let me tell you from bitter experience, final cleaning is *surprisingly easy to underestimate*. The scope and cost of your responsibilities can be substantial, running into thousands of dollars. Common final cleaning includes:

▌ building surfaces, inside and outside

▌ glass, glazing, and mirrors (including scraping the labels off of all the windows)

▌ floor coverings (carpet, resilient, hard tile, etc.)

▌ concrete (or uncovered) floors, which can present difficult cleaning challenges

▌ raking and general cleanup of the outside areas

▌ wiping down surfaces of plumbing, mechanical, and electrical equipment and fixtures

▌ cleaning food service equipment to sanitary condition

▌ providing for rodent or pest inspections and/or services from a qualified exterminator

▌ specific items related to individual projects (check your specification manual)

Removing Temporary Constructions

The general contractor must also make sure all temporary enclosures, fencing or barriers put up during construction are removed. This includes:

▌ removing public safety barriers and walkways

▌ taking down barriers built to protect adjacent properties

▌ removing fencing around the building site

▌ disassembling and removing cold-weather constructions built to retain heat so work can proceed

▌ taking out temporary access roads and restoring the surface to acceptable condition

▌ removing mock-ups or test constructions created during the submittal process stage

▌ disconnecting and removing temporary utilities (such as electrical or phone)

▌ disassembling and removing any temporary fabrications or assemblies that aren't included as part of the owner's final product

Coordinating the Closeout Activities of Subs and Suppliers

As a prime or general contractor, you have your own set of closeout activities that you must perform. But in addition, you also have to ensure that all your subcontractors and suppliers provide the closeout material required for their scope of work. The absence of just one of their items may *also* stop the money flow, and that's something we all want to prevent. Some notable items in this group include:

▌ balance reports from the HVAC contractor

▌ tagging of service equipment by the electrical contractor

▌ O & M manuals for equipment

▌ testing and start-up documentation of new equipment

These closeout responsibilities for each trade are described in the individual (CSI) trade specification sections of the project manual. If a sub or supplier is slow at producing the needed closeout materials, you can use a project closeout letter like the one in Figure 13-2 to remind them of their closeout responsibilities.

Training for Customer's Employees

If you install a product or piece of equipment for your project that requires continual maintenance or prolonged operation, you may be required to provide some type of formal training for the owner and his employees. These training sessions are generally given by the firms who install the equipment and may involve information and demonstrations such as:

▌ start-up and shut-down procedures

▌ general safety procedures

▌ emergency operations and procedures, including hazardous material cleanup

▌ adjustments for excessive noise and vibration of equipment

▌ proper handling of record documentation and equipment logs

▌ how to read and use the O & M manuals

▌ how and where to order spare parts and upgrades when required

▌ strategies for economical and efficient operation

▌ effective energy utilization (especially beneficial for larger firms)

▌ special software training, such as with facilities management programs

▌ reading and interpreting identification and warning labels

▌ proper use of fuels and lubricants

▌ general training on specific items or parts of equipment that should be commonly understood, such as showing the staff how to program the thermostats

Providing Extra Stock for Customer

In commercial construction, you're often required by the specification to supply the owner with some additional stock or extra product parts for the maintenance or repair of items that may be difficult to obtain or match. Common items that you may need to supply include:

▌ roofing materials (shingles)

▌ acoustical ceiling tiles

▌ floor tiles

▌ paint mixed to the colors used on the project

Usually the items requested are those that run a risk of having colors or patterns change frequently, or of being discontinued.

Changing Locks

When the job is complete, you have to go through the relatively simple procedure of swapping out the *construction cores* (the key cylinders used during the construction process) on all the lockset finish hardware. Sometimes this switch may involve changing the locksets out completely. This is a simple

ABC Contractors

1234 Hammertacker Way, Leinwaiver, WI 53555 • Office (608) 555-1234, Fax (608) 555-1235

Project Closeout Letter

Date: _____   Project Name: _____

To: _____   Project Number: _____

_____   Owner: _____

Attn: _____   Contract for: _____

All items should be delivered to our office no later than _____

Dear Subcontractor/Supplier:

We are in the process of closing out the project listed above. In order for us to complete your portion of the closeout procedure, we will require the following closeout items:

❏ Signed subcontractor agreement   ❏ Operations & maintenance manuals/information

❏ Certificate of insurance   ❏ Warranty information for _____

❏ Punchlist satisfaction   ❏ Lien waiver for (period): _____

❏ As-built drawings   ❏ Invoice for (period): _____

❏ Other closeout documents: _____

_____

**Please forward these items by the above date in order to avoid
any delay in processing of your payment. Thank you.**

File: CLOSLETR

Figure 13-2
Sub and supplier closeout letter

security feature that makes the keys used by the workers during the construction process useless. You then need to set up the new keys and keying systems, including any master keying, keying common locks, grand-mastering, and so on, for the new owners.

Administrative and Accounting Responsibilities

There are also the classic "office work" responsibilities that you must provide at the end of each project. They include:

▌ the preparation and submission of final pay requests, along with supporting documents, such as final lien waivers (see Figure 13-3), releases, or affidavits

▌ request for the release of retainage funds (money withheld from the general and subcontractors as a contingency)

▌ liquidated damages settlements

▌ change order reconciliation, which unfortunately, often occurs late in the game

▌ any other accounting loose ends

These items may be required of you before you receive the coveted *Certification of Substantial Completion* or *Certificate of Final Acceptance* from the architect. This form signifies that you've complied with all the requirements, and that the final payment's on its way. Figure 13-4 is a sample of a Certificate of Substantial Completion. You can obtain copies of this form from the Associated General Contractors of America. Their address is:

Associated General Contractors of America
333 John Carlyle Street, Suite 200
Alexandria, VA 22314
Phone: 1-800-AGC-1767
Fax: (703) 837-5405

Or, you can also order forms from their Web site at:

http://www.agc.org

Other administrative requirements and responsibilities include:

1. coordinating the project closeout with your surety or bonding company and obtaining a consent of surety

2. contacting the new owners regarding the changeover of insurance and utilities, and taking final meter readings on those utilities

3. measuring stored fuel (if applicable), and taking care of any other "changing-of-the-guard" issues to insure a smooth transfer

Miscellaneous Closeout Items

There are many other items that could be required at closeout as well. Since the requirements vary from architect to architect and from owner to owner, the only sure way that you'll know exactly what's required of you is by thoroughly reading your closeout responsibilities in the specification manual. In addition to the items we've already discussed, you could be responsible for turning over project photographs, damage or settlement claims that occurred during construction, property or mortgage surveys, and more. If you need organizational help, you may want to develop a closeout checklist like the one shown in Figure 13-5 to aid you with the process.

Don't Wait Until the End

Here's another tip: Don't wait until the end of the project to get organized. If you keep detailed, consistent, daily and weekly job logs on site, you'll reduce the administrative burden of closeout. You can eliminate the ritual "scrambling around searching for documentation" that often slows down the procedure, especially satisfying questionable punch list items, clarifying last-minute change orders, or clearing up disagreements about what happened way back when. The information in your field records will help you reconcile situations that may have roots dating back six months or more, thereby avoiding disputes and the potential delay they may cause.

ABC Contractors

1234 Hammertacker Way, Lienwaiver, WI 53555 • Office (608) 555-1234, Fax (608) 555-1235

Waiver of Lien

To whom it may concern:

The undersigned has been employed by _____ to furnish

material, labor, & equipment under the _____ contract for the

_____ project located at _____,

in the city of _____ and state of _____

and owned by _____.

Whereas in consideration of $ _____ dollars paid to the undersigned, the undersigned

hereby acknowledges receipt of said amount and in turn waives any and all rights, claims,

and/or ability to file a mechanic's lien against the project described above or any and all

improvements to the project therein.

This waiver has been voluntarily executed by the undersigned with full knowledge of the rights

afforded under the laws of the State of _____ and all

other applicable governing bodies. By signing this document, the undersigned also certifies

that all materials, labor, and subcontractors employed to execute the work above have been

paid in full.

Signed this _____ day of _____, 20 ____.

_____
Company name (please print)

_____
Signature of company representative

File: LIENWV

Figure 13-3
Lien waiver

THE ASSOCIATED GENERAL CONTRACTORS OF AMERICA

CERTIFICATE OF SUBSTANTIAL COMPLETION

Definition of Substantial Completion

The date of Substantial Completion of Work or designated portion thereof is the date when construction is sufficiently complete, in accordance with the Contract Documents, as modified by any Change Orders agreed to by the parties, so that the work or designated portion thereof is available for use by the Owner.

CONTRACTOR: ◆ Project No.: _____ ◆

TO (Owner): ◆ Date of Issuance: _____ ◆

Project or Designated
Area Shall Include

_____ Project: _____ ◆

_____ Address: _____ ◆

_____ Architect/Engineer: _____ ◆

_____ Contract for: _____ ◆

The work under this Contract has been reviewed and found to be substantially complete. The Date of Substantial Completion is hereby established as _____, which is also the date of commencement of warranties ◆ and guarantees required by the Contract Documents.

A list of items to be completed or corrected, prepared by the Contractor or the Architect/Engineer, or both, is appended hereto. Corrections or changes called for in this list will be made within _____ days from the date of this Certificate. Signing ◆ of this Certificate of Substantial Completion by the Owner in no way alters the responsibility of the Contractor to complete all of the work in accordance with the Contract Documents, including untested or deferred work.

Architect: Contractor:

By_____ Date:_____ By _____ Date: _____ ◆

The Owner accepts the Work or designated portion thereof as substantially complete and will assume full possession thereof at _____ (time) on _____ (date). ◆
Any responsibility of the Contractor to provide equipment operation, maintenance, heat, utilities and security under the Contract Documents shall terminate at the stated hour on the stated date.

Date_____ ACCEPTED BY: _____ ◆

OWNER: _____ ◆

_____ ◆
Authorized Representative

AGC DOCUMENT NO. 625 • CERTIFICATE OF SUBSTANTIAL COMPLETION • OCTOBER 1976   11/99
© 1976, The Associated General Contractors of America

Figure 13-4
Certificate of substantial completion

ABC Contractors
1234 Hammertacker Way, Leinwaiver, WI 53555 • Office (608) 555-1234, Fax (608) 555-1235

Project Closeout Checklist

Project name: _____

Owner: _____   Project number: _____

No.	Description	Responsibility	Date complete
1	Punchlist completion		
2	Remove all temporary site facilities, trailers, etc.		
3	Collect all final invoices/billings: complete and deliver final billing		
4	Complete and collect all final lien waivers		
5	Final inspections/secure certificate of occupancy from code body		
6	Contact insurance carrier(s) – policy end		
7	Final cleaning – interior		
8	Final cleaning/rubbish removal – site		
9	Secure as-built drawings (from subcontractors)		
10	Secure operations & maintenance (O & M) manuals		
11	Secure all product warranties (*may be part of O & M)		
12	Systems start-up and customer employee training programs		
13	Supply customer with additional material and spare parts per spec		
14	Collect and transfer keys to owner prior to final keying		
15	Change over utility connections/fees (gas, electric, telephone, etc.)		
16	Reconcile change orders and retainages		
17	Request letter of recommendation from owner		
18	Prepare final owner's manual(s) per specification		
19	Complete final A/E affidavits and closeout documents		
20	Notice of completion/closeout documents from A/E		
21			
22			
23			
24			
25			
26			
27			

File: CLOSLIST

Figure 13-5
Closeout checklist

Your Closeout Success

Of course, in the end, your most important closeout responsibility is to ensure that you've made your customer happy. Successful commercial contracting is fueled heavily by repeat business. If you're ever going to get away from adversarial competitive bids and enjoy more negotiated deals, you need to develop long-term relationships with as many clients as possible. One way to help build those relationships is by sending out a *customer satisfaction survey* at the end of every job. You can use a simple form like that shown in Figure 13-6.

Follow up with your customer to make sure it's filled out and returned. By tracking these responses and acting on customer advice when necessary, you'll be in a better position to serve your clients, especially the repeat client. You'll also be performing a valuable marketing exercise by showing that you genuinely care about your customers and the product you deliver to them.

ABC Contractors

1234 Hammertacker Way, Lienwaiver, WI 53555 • Office (608) 555-1234, Fax (608) 555-1235

Customer Satisfaction Survey

To: _____

Project name: _____

Project number: _____

Date: _____

Dear _____:

Thank you for selecting our company to build your _____. We sincerely hope that you are pleased with the completed project. We appreciate having had the opportunity to work with you on this enterprise, and hope we can work with you again in the future.

In an effort to improve our service, we have developed this brief customer survey. Would you please take a moment to respond to the questions so we can evaluate how we might better serve you in the future? Your comments and suggestions are truly important to us.

		Very satisfied	Satisfied	Dissatisfied	Very dissatisfied	Comments
1	Were you satisfied with our preconstruction services (i.e. estimating, planning, value-engineering, phasing, etc.)?					
2	Were your questions and concerns addressed satisfactorily during the construction period?					
3	Did you find our office staff to be responsive and courteous when you called?					
4	Were accounting and administrative procedures, such as billing and change orders, handled efficiently?					
5	Were you satisfied with the communication between our office staff and our field representatives during construction?					
6	Did you find our field workers to be courteous and helpful during construction?					
7	Were you satisfactorily informed of any changes in the work schedule?					
8	Was the work site kept neat, clean and orderly?					
9	Were you satisfied with the quality of the work in the field (i.e. carpentry craftsmanship, drywall, painting, etc.)?					
10	Were the close-out procedures handled to your satisfaction (i.e. punch list completed, warranties, O&M manuals and as-built drawing provided, final inspections, etc.)?					
11	Did we fulfill our obligations regarding your project to your satisfaction?					
12	Do you feel that you received the value for your money that you had anticipated at the beginning of the project?					

Additional comments or suggestions:

File: SURVEY

Figure 13-6
Customer survey

Chapter 14

Accounting, Collections, and Your Financial Health

There's no need to panic — and no need to skip past this chapter. I know you want to, but don't! We *are* going to talk about accounting procedures, but I'm not going to try to turn you into an accountant. Besides, only accountants can be accountants — they're just born that way.

The Importance of Accounting

It's easy to make fun of accountants and bookkeepers, but we really do need them. Accurate and complete accounting procedures are vital to the success — and even the survival — of any commercial construction business. Over the years, I've seen many contracting companies come and go, some belonging to close friends. The circumstances varied with the individuals, but without fail, each of these sad business statistics had a common thread. All suffered from poor accounting and collection procedures, which played a major role in their downfall.

Cash flow was always a problem for them. They never had enough money on hand to operate effectively, or should I say, to operate *proactively*. They were always one step behind and always on the defensive. They learned, all too quickly, that the checks don't roll in on a regular basis. Whether times are good or bad, some customers pay on time, some pay late, and some don't pay at all. All of these contractors were skilled craftsmen, but poor businessmen.

In our father's time, they might have gotten by. Superior construction skills brought success. But in today's highly-competitive, complex commercial construction world, the owner/contractor who doesn't have a firm grip on his financial situation is doomed to failure. The "robbing Peter to pay Paul" syndrome sabotages daily operations, creates fiscal chaos, and eventually leads to bankruptcy.

If you don't know where you stand financially *every moment*, you can't intelligently assess the position of your company and plan for future growth. If you're a contractor who's so caught up in the day-to-day grind of contracting that you can't plan for the future — you probably won't have one. Let's take a look at the ways you can avoid that fate.

Bookkeeping vs. Accounting

First, let's draw a distinction between *bookkeeping* and *accounting*. Most of what you'll be doing in your office on a daily basis is bookkeeping. The role of the bookkeeper is to handle the routine tasks of recording and entering daily transactions into ledgers or journals. He or she keeps a paper trail of income and expenses, and maintains up-to-date information on the company's cash reserves, current income and expenses, and interim (short-term) profit and loss. When you first start out in business, it's likely that you'll have one person on your staff who'll do most all of the bookkeeping for your office.

The accountant comes into the picture once the bookkeeper's duties are complete. You probably won't be in a financial position when you start out to justify the expense of having a full-time CPA (Certified Public Accountant) on your staff. That role usually goes to an outside accounting firm.

The CPA, using the data gathered and organized by your bookkeeper, will create a financial picture of your company. This involves putting together accounting documents such as financial statements, balance sheets, profit and loss statements, construction-in-progress reports, and more. Once your finances are organized, your accountant can help you identify potential tax liabilities, cash-flow problems, and investment opportunities, as well as advise you on the best time to make big-ticket purchases or expand your company's operations. It's the accountant who helps you develop the long-range plan that most contractors don't have time to come up with on their own. Since you'll more than likely be going to an outside accountant, we'll focus on the bookkeeping end of your accounting procedures in this chapter.

Common Bookkeeping Duties

Now that we know which role the bookkeeper plays and which belongs to the accountant, let's look at the bookkeeper's duties in more detail.

Accounts Payable (A/P)

Accounts payable is recording, tracking and paying what you owe. For a contracting business, this will mostly involve supplier (vendor) and subcontractor invoices. Your bookkeeper will track and release partial or monthly payouts to vendors, withhold retainage amounts (which may vary in percentage), keep up with or request vendor discounts, and apply payouts to their respective jobs, purchase orders, and journals. The bookkeeper also tracks and monitors all subcontractor transactions to guard against overpayments and/or overbilling, and makes sure that sub payments are up-to-date. Change order adjustments and backcharges (charges that general contractors sometimes charge their subs) are recorded and assessed for each subcontractor.

Accounts Receivable (A/R)

Accounts receivable involves collecting the money that's owed to you. This includes generating invoices for the work you perform and then tracking the progress of the payments made to the company. Sometimes this requires additional collection techniques, which we'll discuss shortly. Payments are recorded as they're received, and balance-forward accounting keeps track of less-than-punctual payers and those who contract to pay you over an extended period. You can use a billing statement like the one shown in Figure 14-1 for clients running a balance. If the account runs delinquent, you can assess finance charges which you add to the balance. The conditions under which an account becomes delinquent, as well as the penalties on delinquent accounts, should be included in your original contract.

Billings for Individual Job Contracts

Most commercial construction companies use standard AIA (American Institute of Architects) forms for billing invoices, particularly G702 and G703 billing and continuation sheets. But on small jobs they may use simpler, one-page statements, like the shown in Figure 14-2. You can also use generic progress billing statements, where you bill based on the percentage of work that you've completed on a job, such as the one shown in Figure 14-3.

Whatever type of invoice you use, your bookkeeper has to track and record all the costs and income associated with an individual job under that job title. Materials, labor, equipment, and general requirement line items must be matched to their appropriate subdivisions or categories, and tracked throughout the job. Change orders and time and material work are added as necessary. Finally, the bookkeeper has to continually update the monthly billing statements to reflect any changes in the scope of work.

Job Costing

In commercial construction, tracking job costs is vital. In job costing, you use individual supplier and subcontractor invoices to record and monitor the

ABC Contractors

1234 Hammertacker Way, Lienwaiver, WI 53555 • Office (608) 555-1234, Fax (608) 555-1235

Statement

September 30, 2000
(Date)

Owner/Customer:

City of Lienwaver, WI
123 Main Street
Lienwaiver, Wisconsin

Project name:

Mayor's office renovation
123 Main Street
Lienwaiver, Wisconsin

Contract for:

General construction

Your P.O./project number:

LW342-2564897-01

Date	Transaction/invoice number	Invoice amount	Credit/payments	Balance
		Balance forward:		$43,251.00
8/30/00	Invoice number 00-0123	1,095.00		$44,346.00
9/15/00	Payment - Check number 1526		30,000.00	$14,346.00

File: BALFWD

Figure 14-1
Balance forward statement

ABC Contractors

1234 Hammertacker Way, Lienwaiver, WI 53555 • Office (608) 555-1234, Fax (608) 555-1235

Job Invoice

Date: September 30, 2000

Owner/customer:
Billy Jack Ford Auto Dealers
321 Jefferson Street
Lienwaiver, Wisconsin

Invoice number: 00.32

Project name:
Miscellaneous remodel/office renovation
Billy Jack Ford Auto Dealers
Lienwaiver, Wisconsin

Contract for:
General construction

Your P.O./project number:
N/A

Date	Work description	Labor cost	Material cost	Sub-contractor cost	Equipment rental	P & O (15%)	Total due
8/30/00	Renovate general manager's office	3,445.00	2,664.23		640.00	1,012.38	$7,761.61
9/15/00	Build new partition in auto bay	654.00	396.10			157.52	$1,207.62
	Total						$8,969.23

Terms: Payment is due 30 days from date of invoice. Late payment is subject to a 1-1/2% per annum finance charge assessed from date of invoice.

File: JOBINVCE

Figure 14-2
Simple job invoice

ABC Contractors

1234 Hammertacker Way, Lienwaiver, WI 53555 • Office (608) 555-1234, Fax (608) 555-1235

Percentage Complete Invoice

Date: September 30, 2000

Invoice number: 00.154

Owner/customer:

Billy Jack Ford Auto Dealers
321 Jefferson Street
Lienwaiver, Wisconsin

Project name:

Interior finish work in GM's office
Billy Jack Ford Auto Dealers
Lienwaiver, Wisconsin

Contract for:

General construction

Your P.O./project number:

N/A

Billing period: September 1, 2000 through September 30, 2000

CSI#	Work description	Base amount including change orders	Less previous billings	Amount due this period	Dollar amount of work completed	Percent complete	Balance remaining
6100	Rough carpentry	$3,000.00	$2,100.00	$ 900.00	$3,000.00	100	$ 0.00
6200	Finish carpentry	1,800.00	0.00	250.00	250.00	14	1,550.00
9200	Gypsum board systems (plaster)	2,250.00	800.00	1,200.00	2,000.00	89	250.00
9900	Painting and finishing	1,200.00	0.00	0.00	0.00	0	1,200.00
	Totals	$8,250.00	$2,900.00	$2,350.00	$5,250.00	64%	$3,000.00

Terms: Payment is due 30 days from date of invoice. Late payment is subject to a 1-1/2% per annum finance charge assessed from date of invoice.

File: PERCTINV

Figure 14-3

Percentage complete billing statement

costs for each project by comparing the original estimated cost to the *actual cost* for a particular task. Many bookkeepers use a vendor coding program with numbered abbreviations to enter vendor invoices against their intended work categories. These codes are often set up in some type of CSI format (Division 2 for excavating, 3 for concrete, etc.).

The project manager generally helps the bookkeeper in job costing by coding the vendor tickets with the appropriate number for each project. For instance, when the PM gets an invoice from the concrete supply company for the footing mix poured on a job, the PM records the name of the job and the cost code on the ticket. He may use a code which looks something like 3.100M, which would be 3 for the CSI category followed by M for material. He would code the invoice from the precast supplier, who both manufactured and installed his product, something like 3.400S. Again, the 3 for the CSI category followed by the S for subcontractor. The work done by the carpentry contractor might be coded 6.100L, with the L standing for labor. These codes help the bookkeeper record the correct cost within the proper work category.

Purchase Orders

The project manager may also help the bookkeeper track the in-house purchase orders for materials and subcontract labor. Purchase orders can be crucial to construction progress schedules because many subcontractors and suppliers won't begin to process your request until they've received a *written purchase order* for the work. They also act as a physical record of supplier commitment, should any questions or disputes pop up later on in the job. Figure 14-4 shows a sample purchase order.

Service and Time and Material Billing

You'll sometimes perform work that can't be defined by lump-sum contractual arrangements. This is often referred to as *time and material* work, or *T & M*. Work that falls into this category might be noncontract service calls, small service or maintenance jobs, or any ongoing undefined or fragmented work.

You can also use T & M billing in cases where it simply isn't possible (or prudent) to assign a lump-sum cost to a particular scope of work. If, for instance, you were remodeling an older building which involved excavation on a somewhat suspect building site, the extent of the work would be unknown. There may be subterranean or hidden circumstances that could run up the costs considerably. Time and materials billing is probably the only way the owner of the property will get any work done on that site.

You can also use a coding process for T & M work to categorize the work. This allows for accurate entry of the labor, material, and equipment used for that particular service or item. A sample of a T & M invoice is shown in Figure 14-5.

Checkbook Reconciliation

One of the more routine, but important, duties of the bookkeeper is matching and reconciling your company's bank statement balance to the office checkbook. Although this may sound simple, the reconciliation process can grow quite involved when there are multiple bank accounts in the office. For each account there's accrued interest, service charges, deposits and bank transfers that need to be recorded — in addition to the checks.

Tracking Equipment Cost and Maintenance

It's no secret that your equipment may be one of your single biggest capital outlays. Your bookkeeper needs to keep close track of your company equipment revenue, expenses, leasing arrangements and payments, maintenance reports and schedules, licenses, fees, transportation taxes, and anything else related to the owning of construction equipment. Doing this accurately can be crucial to maintaining your company's overall financial strength.

Inventory

As contractors, we're far more service-oriented than product-oriented, but you may still maintain some company inventory. If you warehouse some materials, the efficient tracking and control of this

Purchase Order

From: **ABC Contractors**	Purchase order number:	00.6985
1234 Hammertacker Way, Lienwaiver, WI 53555	Vendor Code:	bflc14
Office (608) 555-1234, Fax (608) 555-1235	Job name:	Pennington Estate

To:	Ship to:
Boardfeet Lumber Company P. O. Box 123 Janesburg, WI 55545 Attn: Dick Williams	Pennington Estate W4444 Elite Road Pewaukee, WI

P.O. Date	FOB	Ship via:	Per your proposal #:
9/4/00	Jobsite	Your truck	10368

Buyer	Freight	Remarks	Tax	Req'd Date
TS	FOB		Included	11/1/00

Quantity	Items #	Description	Unit	Ext. Cost
7 ea		3/0 x 6/8 - 1 3/4" Mahogany custom single-panel interior door, slab form (to match existing Pennington Estate profile). No milling /prep /bore. Unfinished.	1,600.00	$ 11,200.00
10 ea		2/6 x 6/8 - 1 3/4" "Same"	1,410.00	$ 14,100.00*
1 ea		3/0 x 7/0 - 1 3/4" "Same"	1,200.00	$ 1,200.00
2 ea		2/6 x 7/0 - 1 3/4" "Same"	1,500.00	$ 3,000.00
1 ea		3/6 x 7/0 - 1 3/4" "Same"	1,800.00	$ 1,800.00
1 ea		6/0 x 6/8 - 1 3/4" (Dbl) "Same"	3,050.00	$ 3,050.00
1 ea		6/0 x 7/0 - 1 3/4" (Dbl) "Same"	3,825.00	$ 3,825.00
1 ea		5/4 x 6/8 - 1 3/4" (Dbl) "Same"	2,850.00	$ 2,850.00
1 ea		2/6 x 6/8 - 1 3/4" paintable grade custom 6-panel interior door - slab form (to match existing Pennington Estate profile). No milling /prep /bore. Unfinished.	1,000.00	$ 1,000.00
13 ea		2/6 x 7/0 - 1 3/4" "Same"	1,010.00	$ 13,130.00
4 ea		3/0 x 7/0 - 1 3/4" "Same"	1,050.00	$ 4,200.00
1 ea		4/0 x 7/0 - 1 3/4" "Same"	1,900.00	$ 1,900.00
1 ea		5/0 x 7/0 - 1 3/4" "Same"	1,920.00	$ 1,920.00
			Subtotal	$ 63,175.00
		*Per telephone conversation of 10/1/00 with door representative, doors are to be delivered no later than November 19, 2000.	Tax	$ 3,474.63
			Total	**$ 66,649.63**
		PLEASE REFERENCE THE ABOVE P. O. NUMBER EACH TIME YOU INVOICE FOR THE ABOVE DESCRIBED WORK.		

Authorized Signature: _____

ABC Contractors, S. S. Saucerman

File: PO

Figure 14-4

Sample purchase order

ABC Contractors

1234 Hammertacker Way, Lienwaiver, WI 53555 Office (608) 555-1234, Fax (608) 555-1235

Time and Material Invoice

Date: September 30, 2000

Invoice number: 00.300

Owner/customer:
Billy Jack Ford Auto Dealers
321 Jefferson Street
Lienwaiver, Wisconsin

Project name:
Interior finish work in GM's office
Billy Jack Ford Auto Dealers
Leinwiaver, Wisconsin

Contract for:
General construction

Your P.O./project number:
N/A

Billing period: September 15, 2000 through September 29, 2000

Item #	Work/material description	Quantity	Unit	Cost per unit	Line total
1	Labor — remove & reconstruct wall in break room	48	MH	34.00	$1,632.00
2	2 x 4 — spf pc studs	22	ea	2.85	62.70
3	2 x 4 — spf pc (plate material)	124	LF	.32	39.68
4	Nails/fasteners	1	LS	26.35	26.35
	Total				$1,760.73

Terms: Payment is due 30 days from date of invoice. Late payment is subject to a 1-1/2% per annum finance charge assessed from date of invoice.

File: TMINVCE

Figure 14-5
T & M invoice

inventory is important for your tax liability. That can affect the profit/loss report for your company. Your bookkeeper must keep close records of all the inventory on hand, track its use, and assign it to the particular job where it belongs.

Loan Amortization

Unless you're independently wealthy when you start off, you'll likely have some type of bank loan or line of credit established to finance your operation. Your bookkeeper should track and make payments on your loans, calculating and adjusting loan payment schedules as necessary. The bookkeeper may also assist you in planning for possible refinancing or restructuring opportunities. Loan amortization is also important because of its relationship to cash flow (which we'll discuss shortly) and your tax liability.

Payroll

Many commercial construction offices have in-house payroll operations. The alternative is to hire an outside payroll company to do payroll for you. In the beginning, you may find this outside method helpful. There are many good payroll companies to choose from, and using one means you'll have one less thing to worry about as you start up operations.

Generating payroll in a commercial construction business can be quite complicated. It may involve a number of responsibilities, such as:

▌ government reporting

▌ multi-state or multi-jurisdictional regulations

▌ certified payroll requirements (sometimes required by project specs)

▌ union requirements (which may include fringe benefits, health and welfare, training, dues, etc.)

▌ applications and fees

▌ workers' compensation

▌ multiple pay rates and scales

▌ multiple departments within your organization

▌ special deductions, such as for a 401(k) or other retirement program

▌ wage garnishment for child support or other legal attachment

Time Sheet Data Entry

Your bookkeeper's payroll duties include tracking the time sheet data. Field workers must turn in their time sheets, usually weekly, so their hours can be entered into the accounting system as job-costing and payroll items. Though this sounds easy enough, this exercise can become a monumental task, especially as your workforce grows and workers are spread out among several different work sites. The bookkeeper has to break down the time sheets, or time cards, by worker, by job name, by task performed, and by the number of manhours required to complete the task. This data is available to be analyzed later for job costing as well as estimating future work.

Cash Flow

Now, let's get to the heart of the matter — money. Making money is what it's all about. Your *cash flow* is vital to your commercial contracting success. While most of your profit (or lack of it) can be traced back to your estimating and project management, effective cash flow strategies can also generate profit for the company.

Efficient cash flow can:

▌ increase your business's interest income

▌ reduce company debt

▌ provide for rainy-day emergencies

▌ fund big-ticket purchases that are often pivotal for your company's growth

▌ elevate your bonding capacity and thus your opportunities for work

So how do you keep that cash flow flowing? Let me suggest a few ways you can do it.

Invoice Promptly

Prepare and send invoices regularly, when they're due. It's easy to get complacent about invoicing. You may think that your time is better spent on something else — but don't fall into that trap. Invoicing is the life blood of cash flow, and lost days and weeks can really add up by the end of the year. If you don't have time to keep up with it, get someone on staff who will.

Spell Out Your Payment Procedures

When you begin a new job, always take the time to explain and discuss your payment procedures with the prospective client. The payment policy is often established in the specification manual, especially on competitive bid jobs. In that case, you may not have anything to say in the matter — you're the one who has to follow the payment procedures. Regardless, a clear understanding of what's expected by all parties is essential to avoiding misunderstandings and disputes over payments. It's surprising how often the participants are totally unaware of something as simple as the date when billings are to be submitted.

Establish a Clear Collection Policy

Be consistent in your collection policies for late payments. It doesn't hurt to diplomatically make your customers aware that you will aggressively go after slow payers. A simple warning can work wonders for nipping slow payers in the bud, possibly preventing a sticky situation later.

Reduce Your Debt

Pay down any credit balance lines and long-term loans as soon as you have the cash on hand to do it. Accelerating your payment schedule cuts down on interest expense, which equates to real revenue for your business in the long run. Besides prepaying loans, make sure that all of your debt is financed at the best available interest rates and terms. If a better rate is available, check into refinancing; a half-point can make a big difference over the life of a loan.

Deposit Promptly

Deposit payments and other income as soon as they're received. The small effort to make timely deposits of payments can add up to big interest dividends and greatly increased cash flow over the course of time. When you consider the huge dollar volume that's commonly involved in construction contracts, it doesn't take a math genius to see that there are potentially large interest amounts at stake. Even though each deposit may not be enormous, they all add up. Depending on how quickly the money is drawn out, the balance in the account can remain quite high for an extended period, and that can earn you an unexpected bonus. Why not take advantage of that "found" money?

Also, check into a sweep account for your checking. It can sweep the balance in your checking account into an interest-bearing account on a daily basis. You may be required to keep a minimum balance, and you may have to ask for it specifically. But in the long run, it will help you to maximize your interest income and minimize your bank charges.

Use Discounts

Take advantage of fast-pay discounts offered by suppliers and subcontractors. Often vendors offer anywhere from a 1 to 3 percent discount, or sometimes even more, on bills that are paid within 10 days. No discount is offered between 10 and 30 days, but after 30 days their finance charges kick in! If you always pay before the 10th day and take advantage of the discount, the savings can add up to a substantial amount by the end of a year. When you consider that your net profit may often be in the 1 to 3 percent range on some jobs, these "minor" savings can do a lot to inflate your bottom line.

Collection Procedures and Overdue Accounts

As mentioned earlier, not all of your customers are going to be good about paying. You'll have slow-payers and no-payers, and occasionally you'll find yourself in the unfortunate position of having to go out and demand your money from clients. Believe me, this isn't unusual, and you're not doing anything wrong. But demanding money isn't easy for most people. It's an acquired skill. To give you a little boost, I'm going to let you in on a secret that has worked well for me.

I've found that one of the best ways to overcome the stage fright involved in demanding payment is to remind yourself that the demands you're making aren't personal — you're simply performing a necessary business function. That function is collecting the money owed to your company so that you can ensure the cash flow that you need for your business to survive.

You don't dislike this person and you don't suspect some hidden agenda. There are no personalities involved! Keep the collection process on a business level, to help prevent any personal animosity. Then you can feel calmer when you have to approach the situation. This attitude is reflected in your tone of voice, and surprisingly, it can make the whole exercise more civil and efficient — and most important — more effective as well. Remember, this is commercial construction. You're not dealing with homeowners who have no understanding of business. All your clients are savvy businesspeople who know full well the importance of collection and cash flow. They know exactly why you're so concerned.

There will, of course, be the occasional tough nut. Unfortunately, there are people who pride themselves on their ability to get by with whatever they can. They'll pay, but at the slowest of paces, and they'll find fault with virtually every aspect of the deal that, up until a short time ago, was perfectly acceptable. Their goal is to force a credit, come up with a creative means of stalling, or find some other evasive tactic. But as a business owner or manager, you have to make the effort.

Occasionally a client may have a justifiable gripe. You or your crews can make a mistake that justifies some type of monetary negotiation with the customer. If you do everything you can to make the client's situation right, but they still refuse to pay, then what? It may be time to bring out the big guns: collection.

Avoiding Conflicts

I've found, however, that it often doesn't need to get to that point. There are less adversarial means that you can employ long before any major conflict ignites. Some of these, if put in place at the very beginning of your relationship, can greatly increase your chances of collecting the money that's rightfully yours. Consider these techniques when you first begin working with a client:

1. Know your customer. This sounds like very basic advice, but many contractors fall short when it comes to investigating the financial situations of their clients. Simply put, the most effective way to reduce past due accounts is to know the people with whom you do business. And "seeming like a nice guy" doesn't cut it! Get past or personal references, a Dun & Bradstreet report, a credit report, or use whatever other credit checking avenues are available to you.

2. Diplomatically let your customers know, early in your relationship, that credit with your company is a privilege . . . *not a right.*

3. Make sure all of your agreements *are in writing!* (This is *really* important! Don't learn this the hard way.)

4. Include wording in your agreements and on your statements that calls for the customer to pay the cost of collection fees in the event of nonpayment.

5. Discourage extended payment terms. Encourage your sales and marketing staff to avoid this option when they're discussing payment arrangements with clients. Sometimes there may simply be no other way to arrange payment. When that's the case, you've got to scrutinize these requests carefully. Stringing out payments over a long period damages your company cash flow. And once a customer has had that arrangement, he'll want the same treatment the next time. Soon, even those who can fully afford to pay up front will be asking for extended payments. Why? To better *their* cash-flow positions!

6. Try to secure personal guarantees from new or smaller client companies that aren't able to, or won't, furnish adequate financial or credit information to you. If they won't cooperate in this, it may be advisable to pass on the job.

7. Try to get a down payment from the client when possible and appropriate. Again, this may be something you do with newer or untested clients. If you don't get a down payment, make some arrangements in the contract for a minimum carrying charge to cover the "cost of money" that you have to put up to get the job going.

8. Always keep the lines of communication open with the customer. As long as you're both talking, there's always hope of working problems out. Also, make sure that you always get through to the right person regarding payments on the account — the one who makes the decisions.

9. When going after past due balances, always attempt to secure at least a partial payment. These payments show "good faith," and that's generally a positive sign that the customer will own up to his obligation. Besides, *some* payment is always better than *no* payment.

10. If you find yourself in a dispute with a client, attempt to resolve it as quickly as possible. If the client's complaint is about poor workmanship, get the work corrected immediately or explain to the client why it doesn't need to be corrected. As I said earlier, a customer will sometimes use a minor dispute as a reason for not releasing a payment. In this case, demand that the undisputed portion of the payment be paid immediately, and tell them you'll negotiate the rest. Then, hold your ground. Be pleasant, but firm — and remain focused in your conversations. Never threaten: just warn them of the possible consequences of nonpayment, and always be prepared to follow through with what you start.

Writing Collection Letters

Most of the time, when customers are notified that their payment is expected immediately, they'll own up to their responsibilities and pay their bill. There may be a reason for the lateness. It could be a simple oversight, or maybe some extraordinary circumstance. First give them the benefit of the doubt. Then, assuming nothing extraordinary occurred, but your best *verbal* diplomatic efforts fail, or you can't even reach the customer by phone, it may be time to mail or fax a collection letter.

A collection letter does a couple of things:

1. It puts the demand in writing (which may be valuable later in case of dispute).

2. It provides the customer with a less personal means of responding than a sometimes very defensive telephone conversation. That may give them the opening they're waiting for.

Most people really don't want to default on their debts. The collection letter gives them *one more chance* to make things right before the situation becomes adversarial.

When writing the letter, it's important to remember that there's always the possibility of a simple misunderstanding. Until you know better, try to avoid alienating the customer while advising them that their payment is overdue. You can, if need be, use a series of collection letters to request overdue payments, each letter growing more firm in tone.

Here's a sample of the phrasing you may want to incorporate into your letter:

"We enjoyed the opportunity to work with you on your recent building project, and we look forward to working with you again in the future."

"Enclosed is an invoice for the work that we have now completed, along with a mailing envelope for your use. Our accounting system alerts us when accounts are not paid within ___ days. We're sure it's just an oversight, and we'll be looking for your quick response."

"If there are special or extenuating circumstances why this account has not been paid, please contact our office at (555) 333-3333, and perhaps we can discuss other arrangements."

"Our company policy requires us to add a penalty of __% to the balance of all accounts not paid within ___ days. We'd like to avoid this charge. Please send your payment by _____ (include a definite date)."

As the letters progress, the wording should include fewer "niceties" and rely on firmer wording to encourage a response. You can say something like:

"This is a second reminder that your account is now ___ days past due. Please remit payment as soon as possible. Thank you."

When you get to the end of your patience, the wording would be more like this:

"Unless we receive payment by _____ (pick a firm date), we will be forced to turn your account over to a collection agency. To avoid this unfortunate action, please take care of your overdue balance immediately."

Sending the Account Out for Collection

If, after repeated tries, you still haven't received a response, you may have no other recourse but to turn the account over to a collection agency. You'll normally only resort to this measure after you've tried every other means of collection available to you. At best, you'll only get a portion of what's due you when it goes out to collection. But that's better than nothing at all. By the way, never threaten collection and *then not follow through*. If you do that, you might as well kiss your money good-bye.

How do you know when it's time to bring in the collection agency? Well, here are some indicators:

▌ The debt is 120 days overdue.

▌ The office time you've expended simply doesn't justify putting any more in-house time into collection efforts.

▌ The customer lies, shows bad faith, or loses credibility by some other means.

▌ Your investigation uncovers circumstances in your customer's financial situation that indicate they'll have difficulty paying, they have other creditor collections in the works, or they're involved in abnormal or possibly illegal business dealings.

You can try bringing a lawsuit against your client to collect money due you, but most seasoned businesspeople would probably agree that unless you're dealing with a substantial amount of money, suing seldom makes good economic or business sense. The sad truth is that it simply isn't worth the legal fees, the lost time, or the aggravation.

Computers

The accounting and bookkeeping end of your business contains endless challenges. Thank heaven for computers and the many readily-available financial and accounting software programs that are specifically designed for the commercial construction office. They help to make our trials a little bit easier.

For example, you can get programs that allow the user to interface job costing with accounts payable and receivable; tie in your purchase order program with your subcontractor control sheets; hold your original spreadsheet estimate in storage for comparison throughout the job, and much, much more. Then of course, there's automated report and check printing, database merge files for mailing bills, direct online services for conducting banking transactions, and so on. In short, there's really a lot of help out there for you. All you have to do is access it. One of the programs that I've found to be especially good is *QuickBooks Pro* by Intuit. It combines all the features I need in a program — and it's easier to use than any of the other accounting programs I've tried. Craftsman Book Co., the publisher of this book, offers a guide to setting up QuickBooks especially for a construction company. Check the order form in the back of the book.

When it comes to accounting, collection and your financial situation, we've only just touched the tip of the iceberg. You'll almost certainly want to retain the assistance of a competent accountant or accounting firm to help you with the more complicated accounting and financial reports. They can organize your finances and set you on the right path, as well as help you with future plans and strategies that'll give your company direction.

Chapter 15

Bonding, Insurance, Workers' Compensation, and Safety

"Richard is the general contractor on the museum renovation project. Anne, the owner's rep for the project, tried to reach him all morning, but no one answered at his office. Thinking that was strange, she tried a couple more times. But each time the phone just kept ringing. The feeling that something was very wrong had been coming over her for the last few days. Not only was no one answering the phone this morning, but Richard's crew hadn't shown up for work last Friday, or Monday, or Tuesday. His secretary had put Anne off for the last few days, telling her Richard was out and would call her when he returned. But he didn't call, and that was unusual, because he had always returned her calls in the past. And now *no one* is answering the phone.

"Anne had heard a few rumors lately — but there are *always* rumors floating around. Besides, she checked him out. Richard's company had a good reputation, and he seemed like a real straight-shooter. The architect had also carefully prescreened all of the bidders. He had insisted on doing everything by the book. But here it is, Wednesday morning, and Anne is facing the all-too-real possibility that her project, which is only 80 percent complete, is about to come to an abrupt halt.

"The fact is, Richard's company went out of business today.

"All the work that she's put into this project, all her plans for the grand opening, all the people she's had to solicit for funds — the thoughts reel through Anne's head. *What about the grand opening? The date's already set! The fliers have been printed and sent out. What am I going to tell all of those private donors who contributed to the project? Where do we find another contractor who'll come in and finish the project at this point? Where do we stand legally? Are we going to be sued? Are we going to have to sue R & R Construction?"* So many considerations; she doesn't know quite where to begin.

"Richard is also in shock. R & R Construction, Inc. had been a well-respected general contracting company for over 40 years. Richard's father started the company in 1957 and Richard took over as president when his father's health began to fail 13 years ago. But this morning, the company closed forever. How is he going to tell his father, and his clients — especially Anne, that nice lady at the museum, who's worked so hard to put that project together? He didn't understand it himself. As in the case of many failures, there didn't seem to be any clear, explainable, reason for R & R Construction's sudden demise. They'd had a few bad years — but everybody does. There were a few failed investments, like the two spec homes out in Timber Ridge. They were real losers. There was also that build-lease fiasco with that out-of-town paper manufacturer. Then, of course, there was the First National Bank project. The bank was the final nail in his company's coffin.

"But it still didn't make sense to Richard. He'd done his homework. All the numbers had worked on paper; things just got out hand. He'd put his own money into that project. He thought he was investing in a bank that held such promise — and he thought he was fortunate to have such a willing and brilliant partner. At least, the guy had seemed brilliant at the time. He doesn't see his partner much anymore — not since the bank failed. In fact, they're not partners any more, and they only speak to one another through their lawyers. *"It'll take years to straighten out this mess,"* Richard's lawyer told him. *"Bankruptcy is your only option. Your debt is astronomical, and all your money's gone."* And after 40 years, R & R Construction, Inc. is gone as well."

Business Failures

When a commercial contractor defaults on a building project, there are no winners. The contractor loses or damages his livelihood, the project's owner finds his investment in dire jeopardy, and all the construction workers are abruptly out of work (and there's a good chance that their last paycheck bounced as well)! The shock waves set off by the default reverberate down the line, inflicting harm on everyone involved in the construction process, including subcontractors, suppliers, and lenders.

By its very nature, construction is a risky business. According to the AGC (Associated General Contractors), half of all the construction firms in business today will be out of business six years from now. That's a sobering statistic. Dun & Bradstreet's *Business Failure Record* reports that 80,000 construction firms failed between 1990 and 1997, piling up a total liability of almost $22 billion. If you've been in this business for any time at all, you probably know at least one contractor, or someone who's done business with a contractor, who's had firsthand experience with a construction business in default.

Bonding Companies

With so much potential for failure in the construction industry, it's no wonder that there's a need for *bonding companies*. The fact that contractors come and go hasn't escaped the attention of business people looking for stable commercial construction companies to deal with. These clients are often successful community or business leaders — and they didn't get where they are by being careless with their investments! Since a commercial construction project is a major investment, no responsible business person is going to take a gamble on a builder going belly-up in the middle of his project. There's simply too much at risk. So, the bonding company enters the picture.

Bonding companies, sometimes called *surety companies* or just *sureties*, offer bonds that provide security because the surety company will back up the commitment that the contractor made to the owner if the contractor won't or can't:

1. perform the work for the cost they have quoted, or

2. pay his workers, subcontractors, and suppliers appropriately and punctually, according to the original agreement.

The surety company uses its financial resources to finish the job and sort out the mess if the contractor defaults. The bonds also protect the owners from sub or supplier liens that could result from nonpayment. The contractor (or actually the owner, since this cost is usually passed on through the contractor's bid) pays the surety company a fee for this service. It's just like paying an insurance premium. When a commercial contractor is first starting out, the fees for a bond can be substantial. But as the contractor gains experience and a sound financial history, the fees go down. Bonding companies periodically review and reassess the fees they charge in relation to the risk they perceive they take with each particular company.

Types of Bonds

There are generally three types of bonds used by the commercial contractor. They are:

1. Bid bonds

2. Performance bonds

3. Payment bonds

The *bid bond* provides financial assurance to the owner that the contractor's bid *was submitted in good faith* and that the contractor intends to enter into the contract at the price quoted at the time of the bond submission. The bid bond may also insure that the bidder will provide the owner with performance and payment bonds, if required, in a timely fashion when the agreement is sealed.

The *performance bond* protects the project owners from financial loss should the contractor fail for any reason to perform the work as set forth in the contract documents.

The *payment bond*, sometimes referred to as the *material payment bond*, insures that suppliers are paid according to the project agreements.

For private construction projects (those not using government dollars), the owner has the option to bond the project or not. An owner who has worked with a particular contractor for many years, and on many projects, may feel confident and comfortable enough with their business association to waive any bonding requirements. This usually only occurs when considerable trust and respect have been built up over the years. Of course, this leaves the owner open to a very messy situation should the contractor default, but this risk is somewhat offset by the cost savings that result from eliminating the bond requirement.

In federal public work, bonding isn't an option. The *Miller Act of 1935* requires performance and payment bonds on all public work projects in excess of $100,000. In addition, state or local mandates often require bonding for certain public projects, or projects that involve any government financing. You can check Division 1, *General Requirements*, of the specification manual to find out whether or not bonding is required on projects you're considering. If you can't find the information there, or if the wording is ambiguous, contact the architect or owner.

Obtaining a Bond

You may already know of a good bonding company through your acquaintances or associates in the business. If not, ask around. It won't take long to develop a "short list" of companies you can call. Schedule an appointment with one, or more, to discuss your situation. The surety company will begin the process of prequalifying you and your company. This means that they'll do an in-depth investigation into your work history, financial status, credit experience, and personal background. It's a very thorough process, and before they'll issue a bond, the surety will weigh and assess these and other questions:

▌ Have you proven, through your company's work record, to be responsible, reliable, reputable, and of good character? (Yes! Your character *does* matter!)

▌ When and how was your company formed? Did you purchase the business from someone else, was it passed down in your family, or did you start it yourself?

▌ How much money did you start with, and how much do you have now? If you're not the original owner, how did past operations compare with current ones?

▌ Do you have proven experience performing the type of work you'll be bidding?

▌ Do you have, or are you able to obtain in an acceptable manner, the equipment necessary to perform the type of work in question?

▌ Do you have the financial strength to support the level of work you want to perform?

▌ Do you have a good credit history and payment record?

▌ Do you have an established line of credit and solid relationship with a reputable banking institution?

Determining the Stability of Your Company

As the relationship between you and your surety grows, they'll start also using other methods for gauging the stability of your company. However, your financial strength will always be your most important asset. The surety will periodically make assessments about your financial situation using *ratios* and *trends* that they've adapted to our industry. The numbers may vary from surety company to surety company, but in general, they all look at about the same criteria.

Current Ratio

The surety will attempt to determine your *current ratio*, which indicates the liquidity of your company. You find your current ratio by dividing your *current assets* by your *current liabilities*. Generally, a current ratio of more than 1.3 to 1 is considered acceptable for the construction industry. This measure of liquidity is crucial because it directly relates to how well the operations of your company are financed. It's also an indicator of your ability to pay your bills consistently and on time. If your ratio is unacceptable (or in peril of becoming unacceptable), the surety will often work with you to develop a plan to help steer your company's finances more towards the industry norms.

Debt/Equity Ratio

Next, they'll look at your *debt/equity ratio*. The *higher* your debt-to-equity, the *weaker* your financial position as a contractor. Generally, a debt-to-equity ratio of about 2.5 to 1 is considered acceptable, but this can vary depending on other financial factors within your company. If your long-term debt levels are disproportionately high, it indicates that your company's assets are going to be tied up for many years. This means that you'll really have to hustle to come up with the level of continuous work that you'll need to survive. On the other hand, if you have too much short-term debt, it may indicate that you've taken on too much work. This is also a dangerous financial position to be in.

Billing Problems

They'll examine any billing problems that you may have had recently, such as substantial underbilling. If the underbilling reflects the fact that you underestimated the percentage complete for work during a billing period, that means you may have created cash-flow problems for your company through sloppy estimating. This is a problem, but it may be a sign of something even more serious. Underbilling is often an indicator of potential problems or disputes involving unapproved or unanticipated change orders on a project.

The surety company may also check your books for unusual items in your current assets and liabilities. In short, they're not going to be thrilled if they see the new bathroom addition in your house billed as one of your company's past projects!

Past Jobs and Backlog Documentation

They'll review your past jobs and backlog documentation. This is a good indicator of your ability to maintain profits and cash flow in the future. They'll also take a look at your credit and collection policies and offer ways to restore or tighten procedures if they think it's necessary.

Becoming Bond-Worthy

Getting a bond is certainly no cakewalk, and often a young or inexperienced contractor will face a considerable uphill climb when it comes to convincing a bonding company that he's bond-worthy. You may not make it right away, but you need to keep trying. The ability to get a bond can have a marked effect on your business. If you're not considered bond-worthy, you'll be limited in the number and type of projects that you're able to bid. You might find yourself caught in the frustration of a classic Catch-22 situation: "How can I ever get the experience and credit history that's required by the surety if I can't bid on half the jobs out there without providing a bond?"

The answer to that question isn't easy. You'll probably have to continue taking on residential or light commercial jobs until you've beefed up your work history and finances. It takes ambition and hard work to succeed under this system, but these precautions are intended to provide a safety net for both you and your clients. Construction contractors are a bold lot. We have to be. So if a job comes along that looks like it may be a little more than we can handle — we take it anyway! Bonding requirements prevent us from taking on more than we should and getting into situations beyond our experience and financial means. Remember, everyone suffers when the contractor goes under.

If you feel frustrated with the qualification process, it may help to view your situation from the surety's point of view. They may be asked to grant you anywhere from 10 to 20 times your company's net worth in bonding privilege. When you consider the unstable nature of the construction industry, it's a bit easier to understand and accept the surety's need for such a thorough investigation and strict bond requirements.

Default and the Bonding Company

If there's a default, a surety company should step in to perform all or some of the following duties:

▌ Ensure, through the hiring of another contractor, that the bonded project will be completed according to the terms of the contract, and at the original contract price. (Any difference between the original price and final cost would likely be assessed to the defaulting contractor.)

▌ Ensure that laborers, suppliers, and subcontractors are paid according to contract, relieving the owner of any financial loss through nonpayment liens filed against the project.

▌ Expedite the work as quickly as possible to prevent any extended break in the construction schedule. Maintaining the continuity of the work is vital to all concerned.

▌ Quickly take control of the project finances to reduce the possibility of the defaulting contractor diverting any funds for his own benefit — which unfortunately does occur in many cases of default.

▌ Act as an intermediary between the owner and the new contractor in what is almost always a very stressful and confusing period of time.

Insurance

It would be nice if we could simply *trust* one another to always do the right thing. But realistically, we don't live in a world that lets us take that view of contractual business relationships. The bonding and surety process is the only solid assurance an owner has that his project will be completed according to contract should his contractor default. But the commercial contractor also needs other kinds of insurance to cover his company, his employees, and himself.

Contractor All-Risk Insurance

One insurance that you'll want to obtain is *contractor all-risk insurance*. It's known by other titles as well, such as *construction insurance, builder's insurance, homebuilder's insurance* or *master builder insurance*, depending on the insurance company. Regardless of what it's called, there are common elements that you should look for in this type of policy. Most important is that it provides coverage for accidental physical loss or damage at the construction site, or to the property under construction, during the course of the project. You normally purchase a contractor all-risk policy for just the anticipated duration of a construction project, but it can often be extended if the schedule goes beyond the completion date.

This type of policy covers common weather damage, like a partially-constructed masonry wall that's blown over in a wind storm. The insurance company may have the wall rebuilt for you, but more likely, it'll just pay you the cost of rebuilding it yourself according to the limits of the policy. The policy may also cover you from losses incurred when materials or other items are stolen from your job site. They'll probably require that you provide reasonable security (which is likely also required by your work contract), such as fencing in the site and keeping small items locked in a trailer to discourage potential thieves or vandals. This isn't really asking for anything out of the ordinary, because most local codes will probably make you fence off your work area anyway to protect the public.

Your workers' tools probably won't be covered under this policy, but you may be able to cover them under another policy, a *trade policy*.

Trade or Business Insurance

This form of insurance (which may be sold under different names) commonly covers all of those things that the modern construction business owner needs to protect himself from liability, loss, and exposure. This is available to general contractors as well as the contracting trades, electricians, plumbers, sheet metal workers, and so on. It can include coverage in some or all of the following categories:

"You don't need to have an accident for OSHA to come onto your job site."

- ▌ general property
- ▌ public liability (including product liability)
- ▌ personal accident and/or illness
- ▌ tax audit
- ▌ motor vehicle or commercial fleet

The insurance carrier may also be able to provide special coverage for unusual situations. You can only find this out by asking.

A trade policy commonly protects your tools and property, whether the loss is incurred in transit, in a fire, flood (or other natural cause), in a vehicle accident, or as the result of someone forcing their way into your car, truck, or residence. Even mobile phones can be included in this coverage, as well as other personal items in your vehicle, but generally only if they're stolen. You're often limited to a set dollar amount on any one claim, such as $250.

You may be covered for accidents or illnesses that occur outside of business hours, *if* you specifically ask to have this included in the policy. Even property in your temporary care that's owned by someone else can be covered. You can often spread out the cost of the premiums over the year, on a monthly or quarterly basis.

Professional Indemnity

If you plan on doing a lot of in-house architectural or engineering design work, you may want to consider *professional indemnity insurance*. This type of policy protects you and your employees against claims that could result from a negligent act, error, or omission, such as violating a set-back ordinance or not providing proper entrance/egress dimensions for the ADA. The coverage may cover the cost of any settlement, as well as costs incurred in defending yourself, your employees, or your company from the claim. The premium for this package will usually depend on the results of an appraisal of your company by the insurance company (or an underwriter for a professional indemnity policy). They'll assess your company, your experience and your partner (if you have one), and look at many of the same business and financial elements that bonding companies consider. The premium varies with the amount of coverage you want to buy.

Safety and Workers' Compensation Insurance

For most commercial contracting firms, the premiums for workers' compensation insurance are a significant portion of the company's business expenses. The cost for maintaining workers' comp is directly related to the contractor's cost of payroll; the higher the payroll, the higher the premium. The type of work your employees are involved in also affects the cost of the premium. The greater the hazard involved in the work, the higher the premium — so it costs more to maintain workers' comp for roofers than for office staff. The premium is also affected by your company's safety record. The longer you go without a claim or major accident, the lower your premiums. It makes sense to have a sound and effective safety program in place. Not only will it lower your workers' comp premiums, but safety on the job site benefits the employer and employees in a number of ways.

Safety Programs

First, of course, there are the workers themselves. Obviously, no employer ever wants to put an employee at risk unnecessarily, or have a worker injured or killed on the job site. What if a worker did die because of an accident on your site? And what if it was due to your negligence, or even just plain ignorance? Besides the legal problems (which could last for years), you'd have to live the rest of your life with the guilt.

But that's not all. After the accident, there would almost certainly be an investigation and an OSHA (Occupational Safety & Health Administration) review of the matter. The investigation could result in substantial fines or other penalties if there were safety violations that contributed to the accident.

You don't need to have an accident for OSHA to come onto your job site. With some limitations, you can be fined or otherwise assessed at any time for a variety of safety violations. And those penalties can be steep. To protect yourself and your employees, you should put a complete, consistent, and disciplined safety program in place.

There's a lot of good information out there to help you accomplish this goal. OSHA publishes many excellent publications on construction job site safety. You can also get guides from many building associations, such as *Occupational Safety & Health Standards for Construction* from the AGC. If you'd like to contact them, their address is:

The Associated General Contractors
  of America (AGC)
333 John Carlyle Street, Suite 200
Alexandria, VA 22314
Phone: (703) 548-3118
        (800) 242-1767 for publications
Fax: (703) 837-5405
E-mail: *www.agc.org*

There are also many private safety firms and consultants who can set up a program for you and your company. These programs often include:

- OSHA training sessions that lead to certifications

- Instructions on the proper way to fill out and submit accident reports (Figure 15-1)

- The establishment of HAZCOM programs to show how to properly handle hazardous products commonly used on the job site

- Instruction in job site safety protocol (Figure 15-2 is a job site safety checklist)

- The establishment of color-coding programs for items like electrical cords to ensure that they've been inspected for safety and soundness

- Training and instruction in railing and fall protection techniques

- Instructions on the proper procedure for collecting, filing, and using Material Safety & Data Sheets (MSDS)

- A safety agreement, like the one in Figure 15-3, that you can use with workers and subcontractors

You'll also want to conduct regularly-scheduled safety meetings at the job site to keep everyone up to date on safety practices. These meetings, sometimes called "Safety Toolbox Talks," are often an important part of your safety program. A sample safety meeting report is shown in Figure 15-4. You may be able to get some instruction or assistance with these meetings from your local building organizations. Some organizations even have an expert who'll come out and conduct the meetings with you.

Workers' Compensation

The commercial contractor is required by law to carry workers' compensation insurance on his employees. The cost to carry this coverage can be substantial — and it's not likely to come down anytime soon. Rising medical costs, increasing

EMPLOYER'S FIRST REPORT OF INJURY OR DISEASE

An employer subject to the provisions of ch. 102, Wis. Stats., shall, within one day after the death of an employe due to a compensable injury, report the death to the Department of Workforce Development (DWD) and to the employer's insurance carrier, if insured. In cases of permanent disability or where temporary disability results beyond the 3-day waiting period, an insured employer shall also notify its insurance carrier of a compensable injury or illness within 7 days after the injury or beginning of a disability from occupational disease related to the employe's compensable injury.

Insurance carriers and self-insured employers must report all relevant information on this form for all compensable claims to DWD within 14 days of the date of injury.

Personal information you provide may be used for secondary purposes [Privacy Law, s. 15.04(1)(m)]. **See instructions for completing this form on reverse side.**

Department of Workforce Development
Worker's Compensation Division
201 E. Washington Avenue, Rm. 161
P.O. Box 7901
Madison, WI 53707-7901
Telephone: (608) 266-1340
http://www.dwd.state.wi.us/WC

EMPLOYEE

| Employe Name (First, Middle, Last) | Social Security Number | Sex ☐ M ☐ F | Employe Home Telephone No. ( ) |

| Employe Street Address | City | State | Zip Code | Occupation |

| Birthdate Mo. Day Year | Date of Hire | County and State where accident or exposure occurred |

EMPLOYER

| Employer Name | WI Unemployment Insurance Account No. | Self- Insured? ☐ Yes ☐ No | Nature of Business (specific product) |

| Employer Mailing Address | City | State | Zip Code | Employer FEIN: |

| Name of Worker's Compensation Insurance Co. or Self-Insured Employer | Insurer FEIN: |

| Name and Address of Third Party Administrator (TPA) used by the Insurance Company or Self-Insured Employer. | TPA FEIN: |

WAGE INFORMATION

| Wage at Time of Injury $ | Specify per hr., wk., mo., yr., etc. | In Addition to Wages, Check Box(es) if Employe Received: ☐ Meals ☐ Room ☐ Tips | No. of Meals/wk. _____ No. of Days/wk. _____ Avg. Weekly Amt. $ _____ |

Is worker paid for overtime? ☐ Yes ☐ No   If yes, after how many hours of work per week? _____

	Start Time	Hrs. Per Day	Hrs. Per Wk.	Days Per Wk.	For the 52 week period prior to the date the injury report below the number of weeks worked in the same kind of work, and the total wages, salary, commission and bonus or premium earned for such weeks.
Employe's Work Schedule when Injured					
Employe's Normal Full-time Schedule for Injured's Work					No. of Wks. / Gross Amount Excluding Tips $_____ / If Piece Work - No. of Hrs. excluding overtime

| Part-Time Employment Information: | Scheduled Hrs. Per Week | Are there other part-time workers doing the same work with the same schedule? ☐ Yes ☐ No   If yes, how many? _____ | Number of full-time employes doing the same type of work. |

INJURY INFORMATION

| Injury Date Mo. Day Yr | Time of Injury ____AM ____PM | Last Day Worked Mo. Day Yr | Date Employer Notified Mo. Day Yr | ☐ Date Returned to Work ☐ Estimated Date of Return | Mo. Day Yr |

| Did injury cause death? ☐ Yes ☐ No | Was this a lost time or other compensable injury? ☐ Yes ☐ No If no, insurer does not submit report to DWD | Did injury occur because of: ☐ Substance Abuse | Failure to Use ☐ Safety Devices | Failure to ☐ Obey Rules |

| Date of Death Mo. Day Yr | Name, Relationship, and Address of Closest Dependent of Deceased if Injury Caused Death |

| Name of Witnesses |

| Name and Address of Treating Practitioner and Hospital |

Injury Description -What happened to cause this injury or illness? Describe the employe's activities when the injury or illness occurred with details of how the event or exposure occurred. Include name(s) of other individuals involved. Specify tools, machinery, objects, chemicals, etc. that were involved in or caused the injury.

| Report Prepared By | Work Phone No. ( ) | Position | | Date Signed |

SEND REPORT IMMEDIATELY — DO NOT WAIT FOR MEDICAL REPORT

WKC-12 (R. 2/98)

Figure 15-1
Accident report (front)

EMPLOYER AND INSURANCE CARRIER INSTRUCTIONS

The employer must complete all relevant sections on this form and submit it to the employer's worker's compensation insurance carrier or third party claim administrator within seven (7) days after the date of a work-related injury which causes permanent or temporary disability resulting in compensation for lost time. The employer's insurance carrier or the third-party claim's administrator may request that this form also be used to immediately report any injury requiring medical treatment, even though it does not involve lost work time.

For any work injury resulting in a **fatality**, the employer must also submit this form directly to the Department of Workforce Development **within 24 hours of the fatality**.

An employer exempt from the duty to insure under s. 102.28, Wis. Stats., and an insurance carrier administering claims for an insured employer are required to submit this form to the Department of Workforce Development within 14 days of the date of work injury.

MANDATORY INFORMATION

In order to accurately administer claims, each of the following sections of this form must be completed. The First Report of Injury will be returned to the sender if the mandatory information is not provided.

Employe Section: Provide all requested information to identify the injured employe. If an employe has multiple dates of employment, the "Date of Hire" is the date the employe was hired for the job on which he or she was injured.

Employer Section: Provide all requested information to identify the injured worker's employer at the time of injury. Provide the name and Federal Employer Identification Number (FEIN) for the insurance carrier or self-insured employer responsible for the worker's compensation expenses for this injury. Also identify the third party claim administrator, if one is used for this claim.

Wage Information Section: Provide the information requested regarding the injured employe's wage and hours worked for the job being performed at the time of injury.

Injury Information Section: Provide information regarding the date and time of injury. Provide a detailed description of the injury, including part of the body injured, the specific nature of the injury (i.e., fracture, strain, concussion, burn, etc.) and the use of any objects or tools (i.e., saw, ladder, vehicle, etc.) that may have caused the injury. Provide the name of the person preparing this report and the telephone number at which they may be reached, if additional information is needed.

Courtesy: Wisconsin Department of Workforce Development — Workers' Compensation Division

Figure 15-1
Accident report (back)

ABC Contractors

1234 Hammertacker Way, Lienwaiver, WI 53555 • Office (608) 555-1234, Fax (608) 555-1235

Safety Checklist

Project name: _____   **Owner:** _____

*This list should be revisited at least once a week during the entire project duration

Item #	Description	Yes (✓)	No (✓)
1	Injury waivers signed by all visitors to site. Hard hats worn by all employees and visitors.		
2	All job site vehicles and equipment in sound operating condition.		
3	Hearing protection worn by all workers in high noise level areas. Respiratory equipment used as required.		
4	Eye/face protection worn by all workers where applicable (i.e. drilling, burning, chipping, grinding, sawing).		
5	Hallways/exits clear of debris and stock. Clear debris/scrap from exterior walks and doorways.		
6	Job site safety railings, guards, staging, netting, etc. in place.		
7	Gloves worn when handling sharp-edged or jagged materials.		
8	All ladders, cables, ropes, hoists, lifting equipment in proper working order and checked daily.		
9	All electrical equipment properly grounded, shielded, and color-coded for identification.		
10	Finger rings, watches, and other cumbersome jewelry not worn by workers (recommended).		
11	Maintain sufficient lighting in all work areas and stairwells.		
12	Proper footwear worn by workers to protect from ankle sprains, puncture wounds, and impact from above.		
13	Proper precaution used against cement burns and other common chemical abrasion.		
14	Job site safety/accident and OSHA postings in place. Warning signage posted near all hazardous work.		
15	Fire extinguishers, telephones, MSDS sheets, and first-aid kits located in places accessible to all workers.		
16	Fuels stored in proper containers. Oxygen/acetylene tanks stored and secured properly.		
17	All power tools/equipment equipped with proper blade guards and pulley housings. Cut-off saws have auto return.		
18	Scaffolding/staging in good condition and erected according to OSHA standards. Railings and toeboards in place.		
19	Fall protection for workers and safety netting to protect people below are in place and secure where applicable.		
20	Hazardous materials stored and contained according to manufacturers' instructions and MSDS information.		
21	All excavations and trenches sloped properly for safe occupation. Material stored at least 2' from edge.		
22	All transported material, equipment, etc., loaded, cinched, and flagged appropriately.		
23	All workers know to report all accidents, injuries, and unsafe job site conditions.		
24	All overhead utility wires identified and shielded as necessary against overhead craning or hoisting.		

File: JOBSAFE

Figure 15-2

Job site safety checklist

ABC Contractors
1234 Hammertacker Way, Lienwaiver, WI 53555 • Office (608) 555-1234, Fax (608) 555-1235

Safety Agreement

Employee name: _____   **Date:** _____

Project name: _____   **Project location:** _____

To whom it may concern:

I, the above named employee, have received a copy of and have been instructed in the _____ safety program, accident prevention program, and HAZCOM program. I have been trained in the use of the fire extinguisher, eye protection, hard hat, ear protection, and other equipment pertinent to my trade. I have been issued a hard hat for the project and will wear it at all times when I am on the project site. If I have an injury, I will report it immediately to the project superintendent or other authority. I have been shown the location of all safety postings, first-aid materials, fire extinguishers, emergency telephone numbers, hospital routes, and HAZCOM information/MSDS sheets for products being used on site. I will report any unsafe conditions to the proper authority and I will not allow unauthorized persons to enter the site without proper clearance and notification to _____. I will wear proper foot protection at all times, and protective gloves when handling jagged or sharp-edged materials. I will maintain my tools and equipment to ensure they are in proper working order, with electrical cords and guards in sound condition, and I will use them only for their proper applications. I will keep my work area clean and orderly to avoid accidents or traffic blockage in case of emergency.

Signed this _____ day of _____, 20 ____.

_____
Name of employee/worker (please print)

_____
Signature of employee/worker

File: SAFEAGR

Figure 15-3
Safety agreement

ABC Contractors

1234 Hammertacker Way, Lienwaiver, WI 53555 • Office (608) 555-1234, Fax (608) 555-1235

Safety Meeting Report

Project name: _____   **Date:** _____

Project location: _____   **Project owner:** _____

Those in attendance (use back of sheet , if necessary):

_____   _____   _____

_____   _____   _____

_____   _____   _____

_____   _____   _____

_____   _____   _____

_____   _____   _____

_____   _____   _____

_____   _____   _____

Safety meeting agenda items (old business) discussed:

_____

_____

_____

_____

_____

_____

_____

_____

New business discussed:

_____

_____

_____

_____

_____

_____

_____

_____

_____

File: SAFEMEET

Figure 15-4
Safety meeting report

litigation of workers' comp claims, and the inability of workers' comp claims management to combat fraudulent cases will only continue to drive premium prices up. To keep the cost in line, you may find it helpful to put a *workers' compensation loss-control program* into place along with your current safety program.

Loss-Control Programs

Your insurance company can help you to set up a loss-control program. The program works to remedy unsafe environments by identifying hazardous site conditions and unsafe work practices. It also attempts to address the human error side of the problem by distributing information, and by training employees on how to deal with common workplace hazards. The insurance company's loss-control specialists can also keep you up-to-date on new or changing OSHA requirements. The establishment of a sound loss-control program can benefit your company by:

▌ lowering your insurance costs

▌ reducing accidents and injuries, and their associated costs, on the job site and in the office

▌ reducing equipment, material, and building damage and/or loss due to unsafe practices

▌ increasing your productivity as the result of fewer accidents

▌ reducing schedule delays and thus increasing profit

▌ improving the quality of work by reducing stress in the work environment

▌ improving employee morale

▌ improving employee loyalty through your concern for their safety and general welfare.

A loss-control program doesn't just deal with prevention. If an employee is injured on the job, the program can also assist with *Back-to-Work* programs to rehabilitate and help the employee get re-established in the work place. Studies have shown that the longer an injured employee stays away from work, *the less likely it will be that he returns at all.* Encouraging a quick recovery and return to work is an essential part of the program. It provides emotional support for the employee by allowing him to return to work, even if it means being on altered or restricted duties for awhile.

Insurance companies that specialize in workers' compensation claims also monitor and help contain spiraling and irregular medical costs by offering these services:

1. a listing of preferred physicians and hospitals

2. a hospital billing review

3. individual case management

4. an extensive database that allows the insurance company to compare your medical bills with the industry standards to determine if they're fair and accurate

With these services available, it's obvious that the insurance company you deal with will play an important role in your ability to control your workers' compensation costs. So, choose wisely and shop around if necessary. Your choice may greatly affect your company profits.

Other Insurance Needs

Depending on your work specialty and your rate of growth, eventually you may find that you have other insurance requirements, such as *health insurance* for your employees and their families. You'll also need to occasionally assess the policies you already have in place. Get together with your insurance carrier to discuss whether they're still effective and if their cost remains competitive on the market. The good thing about discussing insurance is that there's never a shortage of ambitious insurance agents anxious to tell you what's new. Take it all in, and as the information shakes out, you'll know what moves you want to make.

Chapter 16

Marketing and Promotion

You have a plan and you've defined your direction. You know what you want to do. But there's still one key element you have to consider if you're going find success in commercial construction. Without an effective and consistent marketing and promotion program, even the best plan can't carry you through the turbulent ebb and flow of the commercial construction market. Forward momentum and continuity of work are crucial to your success. Without adequate marketing support, the natural downturns in the industry can interrupt your company growth.

And that's not what you want to happen. You want to keep the momentum going! So let's look at a few marketing methods and strategies that you can use to keep your business growing. By using some or all of the ideas we're going to discuss, you can establish a marketing program that'll provide your business plan with the critical support it needs for profitable, long-term sustained growth.

Be sure you establish an adequate budget for marketing and promotion as part of your overhead expenses. When you start out, you may have to devote a larger percentage of your overhead to advertising and promotion than an established firm with the advantage of name recognition and a satisfied customer base. Giving you an exact percentage of what you'll need to spend is a little tricky since every business is different, but I'd say that if you're operating with an overhead of $250,000, you should plan on spending between 2 and 5 percent of your first year's budget on advertising. That would be roughly somewhere between $5,000 and $12,500.

As the business takes off and your sales are closer to your goals, this percentage can be pared back proportionately — although the dollar cost may remain the same or even grow. For example, if your overhead grows to $500,000 in your second or third year, but you're happy with your advertising strategy, you may want to continue with the same spending. But now, $5,000 is only 1 percent of your overall overhead.

Marketing Tools

We're going to discuss all the conventional marketing vehicles, including newspaper advertising, brochures, direct mail, and trade shows. But first, I'll start with what I consider the best commercial construction marketing tool around: *your customer*.

The Greatest Tool of All — Your Customer!

It's true. There's arguably no better advertising than a satisfied customer. And no other business relies more on repeat customers than commercial construction. I have clients for whom I've performed dozens of separate construction projects over the years. Rather than focusing your energies on finding new customers, you'll discover it pays to maintain good relationships with your old customers. All you have to do is treat them right. If you do, you may find that when you finish one job, your next sale is standing right in front of you!

While repeat customers are the heart and soul of a successful commercial contracting business, there are more advantages to working with repeat customers than just finding the next job. When you work with repeat customers, you don't need to start the learning curve at the beginning. You already know them, and they know you. You know how they like to work with people; you have a feel for their moods, working patterns, and their idiosyncrasies. And you know they pay their bills. There are far fewer surprises, and that's a very good thing. The more familiar you are with the client and the scope of work, the better you'll be able to control your operations and costs, and the greater your chances for a good solid profit.

For example, I've had a continuous work relationship with a hospital for over three years now. Most of the work involves renovating patient rooms, staff offices, or special areas such as the pharmacy or chapel. Occasionally, we'll redo an entire wing at one time. We've become quite familiar with hospital construction and the needs of the hospital staff, and have an open-book policy with the hospital reps. With this level of familiarity, we never take a hit on our profit. We go into each project as partners. If prices go up during the construction work, the hospital will absorb the additional costs (to a fair extent — if it goes up too high, some negotiation is in order). In that way our profit is protected. By the same token, if prices go down, the hospital will reap the benefit of the savings. Granted, there are no major windfalls when you work like this, but it's steady, predictable income for our company. That hospital usually provides us with over a million dollars a year in sales volume. Now that's good marketing!

Everybody's Part of the Sale

But the hospital work didn't just happen. As the primary estimator and project manager for my company, I had to sell myself to hospital administration. My field workers had to sell themselves through the quality of their work and their professionalism. And my office staff had to sell themselves by offering prompt, courteous service and efficient accounting and administration. Everyone in the company was part of the sale.

Were there problems? Of course there were. Nothing ever goes entirely right. But the *way we handled those problems showed our true professionalism.* We communicated promptly with the staff, solved the problems quickly, and used what could have been an uncomfortable situation to increase customer loyalty. Although no one likes to deal with customer complaints, sometimes they give you the opportunity to prove that you're sincere in your commitment to your work and your clients.

> *"Your simplest and most effective marketing tool will be your employees' ability to resolve problems or conflicts in a manner that doesn't offend the customer."*

On one of our first jobs with the hospital, for example, a problem came up after the work was completed. We had done a build-out of the clinic in which we redid all the interior space: new walls, ceilings, floors, plumbing, electrical, mechanical, etc. Two weeks after completion, we received a call from the director of engineering at the hospital informing us that the doctors were complaining about how long they had to wait for hot water to flow from the taps in the exam room sinks. The rooms were located about 70 feet from the hot water heater. There should have been a recirculating line from the exam rooms back to the water heater to ensure constant hot water to the rooms, but we hadn't installed one because there wasn't one on the plans.

This was a design-build project, and we had worked with the engineer and the hospital's staff architect to put the plans together. It was an oversight on everyone's part (we are all professionals), so we offered to share the cost of fixing the problem. The hospital was pleased with our attitude — they'd obviously been stuck before — and so the incident, though an error, turned out favorably for us.

The way we handled this problem told the hospital staff more about our company than any brochure could — and we've been working with them ever since.

Think for a moment of a company or service that you especially like. Chances are your reasons for liking or disliking any company have a lot to do with the attitudes, abilities, and service provided by the employees, especially when things don't go quite as planned. In most cases, products and prices are competitive. Service and attitude aren't — and that's what sets you apart. Most customers understand that mistakes occasionally happen. We're all human. It's rarely the mistake itself that causes hard feelings between a customer and a company. Rather, it's the fact that the business doesn't quickly remedy the situation. People want to conduct business where they know they'll be treated well, and they'll forever avoid places where they weren't!

Turn Complaints into Promotion

Your simplest and most effective marketing tool will always be your employees' ability to handle and resolve problems or conflicts in a manner that doesn't offend the customer. I have a few tips for turning customer complaints into good promotion for your company.

Accessibility — Make sure someone from your company is always accessible to the customer. This may sound elementary, but I often walk into contractor's offices and find that there's absolutely no one around. If you depend on a voice-mail system for customer contact, get rid of it! Voice-mail has got to be one of the most ill-conceived inventions ever visited on society. They generally alienate potential customers and infuriate existing ones. *Always have a real person answer your phone.*

Interest — When faced with a customer complaint, always show genuine interest and offer a sincere response. Cavalier or casual attitudes will only further inflame the situation. Listen to the *complete* complaint before you offer any help. Many times, a major part of quelling the anger and resolving the conflict is showing the client that you're interested in his complaint and allowing him to simply "get it off of his chest."

Responsibility — Never make excuses for a problem. Don't blame the workers in the field or the person in accounting. To the customer, you're all one company, not individuals. Passing the blame won't solve the problem. Customers don't care who or why, they just want to know when the problem will be fixed!

Attentiveness — Repeat your understanding of the customer's problem back to him. This shows that you not only understand the problem, but that you understand how important the situation is to him. Let him know that you sincerely want to provide a remedy. Leave him with the feeling that you're on his side, and that you appreciate the fact that he brought the problem to your attention.

Sincerity — Make sure that everyone in your company is open and sincere in their discussion with the customer. Do *not*, under any circumstances, encourage your employees to use silly euphemisms when referring to problems. There's a trend in some companies to forbid their employees from using the word "problem" in conversations with customers. They're supposed to refer to problems as "opportunities." I don't know where this phenomena started (probably at some marketing seminar), but companies that encourage these ridiculous games in dealing with customers are foolish. They're on their way to becoming the subject of a *Dilbert*® cartoon. Remember, your customers aren't stupid. They'll only find this type of behavior condescending and insulting.

Patience — As hard as it may be sometimes, don't get defensive when dealing with a customer. There are some really nasty chaps out there, and they can talk pretty rough. But *you've* got to keep your cool. If you maintain your interest in their problem and stay calm, they'll usually settle down after their first burst of anger passes.

Concern— Here's a good trick that I use all the time. Before going on to discuss a solution for the problem at hand, ask the customer, "Is there anything *else* I can do to help you?" This expression of concern gives the customer the opportunity to unload any remaining frustrations, and it's often the point in the conversation where "the walls come down." Once you've established that you're genuinely concerned about them, and their project, you'll have the customer's confidence.

Action— Try to come up with a mutually-agreeable solution to the customer's problem right away. If a solution isn't possible right away, arrange for a specific first step towards a solution. If you need to check with a sub or another party, set up a time that you'll call the customer back — and then do it. *Always* keep your promise. If you can't get hold of the sub or you can't get the information you need by the time you have to call the customer back, call him back and tell him that. Don't just not call.

Follow Up — Once the problem is fixed, follow up with a call to the customer to make sure everything's done to his satisfaction. If it's not, treat it as a brand new situation and start the whole process over. Of course, you don't need a complaint to follow up with a customer, especially after the job is completed. Make it a point to call or visit a recent project owner "just to see how things are holding together." They'll almost always appreciate it, and it gives you a chance to discuss future work opportunities.

Direct Mail

Direct mail involves sending a brochure, flyer, or other piece of promotional literature directly to a potential client in hopes of attracting their business. It's a fairly simple two-step way to get the word out. First, you can use one of the many desktop publishing and word-processing software packages to create your flyers and brochures. Then use a database program with mail-merge to personalize your message to different clients before mailing them off. The ability to create professional mailers is well within the technical expertise of most computer-literate construction professionals.

But there are other considerations if you want to develop an effective direct mail program. The most important is whether direct mail makes sense for your particular business. Remember, the object of *any* marketing program is to generate revenue for your company. A direct mail campaign can be very expensive. If you've tried a few direct mail attempts and your response rate isn't at least breakeven, you should probably move on to other marketing techniques. In other words, you must make at least as much in profit from the work you get from your mailings as the cost of design and printing, office time and postage for your direct mail brochures. If you don't, then it's either time to rethink the current program (perhaps redesigning your brochure or streamlining your database of potential clients), or consider other ways to advertise.

Figure 16-1 shows part of a mailer that we use for our business. It's simple and interesting — and tells the potential customer some important information about our company as well as showing examples of our work. It's just about the right amount of information for a single page.

We'll discuss the brochure itself in detail shortly. But first, here are a few tips for creating a good direct mail program:

1. Include self-addressed reply cards as part of your brochure or mailing. They'll help to increase your responses.

2. Personalize your mailings. Mention the names of key salespeople, estimators, project managers or foremen in the flyers. It's amazing how many times customers respond simply because they know someone in your organization, perhaps through their church, club, or professional organization. However, *don't overdo it!* Limit the number of mentions. Too many will get in the way of your message.

3. Don't fill your direct mail with information unrelated to the type of customer you're trying to attract. If you're sending out a promotional package on agricultural buildings, stick to that message. A potential agricultural customer won't want to wade through information on residential window and siding replacement. Excess information just increases your postage and development costs, and decreases the chance that someone will take the time to read it. The message should be short, simple, and *focused*.

Gilbank

CONSTRUCTION, INC.

THE GILBANK DIFFERENCE . . . COMPLETE CUSTOMER SATISFACTION

Citizen's State Bank

▌ On-time scheduling . . . no delays and within budget.

▌ Innovative production scheduling accommodates your own day-to-day operations during the construction project.

▌ Safety-minded crews with an old-fashioned work ethic.

▌ Diversified team of professional craftsmen.

▌ Well-rounded fleet of top quality equipment, tools and other resources.

▌ Crews work year-round regardless of the weather.

▌ Ethical approach allows contracts to be completed without unexpected 'extras' that can deplete a budget.

Through close attention to detail, quality, craftsmanship and safety, Gilbank Construction has earned the confidence and respect of numerous repeat customers who have come to expect full service and quality in every project. The reviews are unanimously positive . . . the Gilbank Difference IS complete customer satisfaction.

Sheffield Residence

Lake Geneva Lake Front Stabilization

Gilbank

CONSTRUCTION, INC.

General Contractor

200 Main Street
Gore, WI 55555
Phone: (555) 888-8888
Fax: (555) 888-8889

Figure 16-1
Use pictures and ideas to promote your company

4. Check into bulk-mail permits. They can save you a bundle in postage if you do a lot of mailings. There are some restrictions, including minimum quantities and having the entire mailing the same size and shape, but they're not too hard to work with. There are also bar-coding and presorting options that can reduce your cost even more. In fact, you may already have software that sorts and codes your mailings as part of your word-processing program.

5. Choose an appropriate class of mail for sending out your flyers. First-class costs more, but anything that can't be delivered will be returned to you. That gives you the opportunity to update or correct your mailing list, and could save you money in the long run by eliminating the undeliverable names from future mailings. Also reduce costs by periodically purging your database of old or redundant customer names. There's software available that can help you with this task.

Creating Your Brochure

The brochure is the most popular medium for direct mail advertising. You need to put some time and thought into its design and focus. Having a clear objective or focus for your promotion is essential for creating an effective direct mail brochure.

Ask yourself what you're trying to accomplish with the mailing. I can't tell you how many brochures I get that don't seem to have any discernable message or purpose. I'm sure you've seen them, too. These misguided efforts usually end up in the trash can. You certainly don't want to go to all the expense of a sending out a direct mailer only to have it tossed out because it's not effective. So, let's see if we can help put a little *pizzazz* in your promotion by examining some of the elements of an effective brochure:

▮ Have an objective! I know I'm repeating myself — but it's important! *Target your brochure to your audience.* Don't send out general information. For example, if you do medical construction work, target hospitals, clinics, and doctor or dentist offices. Tell them precisely what you can do to help them, and why your company is the best one for the job. Company histories are nice, but most people (especially the busy executives who make up the commercial contractor's clientele) simply don't have time to read through a lot of fluff. They want to get down to business. It's OK to include a small blurb about your company's excellent people, solid reputation, and company vision, but quickly get on to what your company can do for the reader. Figure 16-2 is a page from the brochure of a firm who specializes in large industrial projects. It promotes the idea of strength, stability and longevity — both in the photograph and the text. It's very effective.

▮ Try to make your brochure visually inviting to the client. Don't have long, continuous paragraphs of information. Break your copy down into small, readable sections with catchy, descriptive titles and subheads. Include photos, graphs, charts, and graphics to keep it appealing. Don't just put your company name as the lead headline. Go for something that will grab their attention.

▮ Use simple, clear phrases and correct grammar. Too much technical jargon is tiresome and confusing to most readers, and will give them a good reason to discard the brochure. On the other hand, don't get too casual, too clever, or use slang. Some people just don't get it, and what you thought was cute could end up being misunderstood — or worse, offensive.

▮ Avoid using tired cliches like *"We're the leader in the industry"* or *"We're number one."* These terms have so saturated our lives that they're almost invisible. People ignore them. Talk as if the person were standing in front of you.

▮ Don't try to pack too much into the brochure. Including information that doesn't underscore your message will only muddy the water.

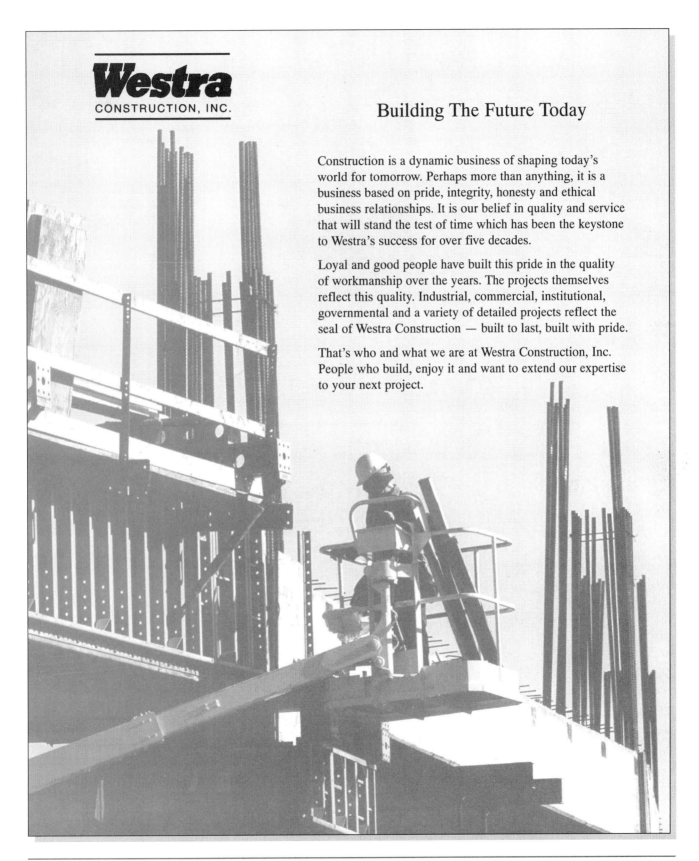

Westra
CONSTRUCTION, INC.

Building The Future Today

Construction is a dynamic business of shaping today's world for tomorrow. Perhaps more than anything, it is a business based on pride, integrity, honesty and ethical business relationships. It is our belief in quality and service that will stand the test of time which has been the keystone to Westra's success for over five decades.

Loyal and good people have built this pride in the quality of workmanship over the years. The projects themselves reflect this quality. Industrial, commercial, institutional, governmental and a variety of detailed projects reflect the seal of Westra Construction — built to last, built with pride.

That's who and what we are at Westra Construction, Inc. People who build, enjoy it and want to extend our expertise to your next project.

Figure 16-2
Brochure promotes idea of strength, stability and longevity

▮ *Ask for the sale.* I know this is going back to "Salesmanship 101," but never forget the importance of a good pitch. Always give the customer an easy way to respond, like "mail in this card" or "call our toll-free number . . ."

▮ On the back of the brochure, include references from past clients and list some of your projects that are relevant to the audience that you're focusing on. Again, if your brochure is meant to appeal to medical clients, include only those related projects.

▮ Choose a heavyweight paper for your mailer. It'll cost you a bit more, but nothing comes off cheaper than a flimsy brochure.

▮ Check out the custom brochure-making software packages that are sold everywhere. They can assist you with structure and visual ideas. Stay away from gimmickry, and avoid using too many colors. With all the bells and whistles in the graphics packages, it's easy to have fun and get carried away. That's something that I've been guilty of, and unfortunately, you only end up distracting the reader from what's most important: *your message!*

▮ Finally, follow up your mailer with a phone call or a visit. A visit is better.

Ads in Newspapers and Trade Papers

When you design an ad for newspapers and trade publications, your approach should be the same as you'd use for creating a catchy brochure. The same rules apply, except with these ads you have to work within a much smaller space! Column inches are expensive. You'll have to limit your message to the absolute essentials:

▮ Go for something visual. Set off your message with a box, or try white on a black background like the ad in Figure 16-3. Make your ad stand out from all the others on the page.

▮ Don't fool around with a frilly headline. Get right to the message and lead the reader through the ad.

▮ Deal in specifics, not general statements. Tell the reader in one or two sentences exactly what you can do for them.

▮ For communicating an idea quickly, nothing beats a picture. Use a photograph of a past job, or perhaps of an industrial park lot. Sometimes you can entice a customer by showing a bare lot and telling them "You could be here!" Others prefer to see a finished project. You can use both — it will help make your ad stand apart.

Customer Presentations

If you're going to be successful in commercial construction, sooner or later you'll have to make a *customer presentation*. These presentations are normally requested when a client is interviewing several contractors for a job he's planning. The goal of these interviews is usually to choose a contractor who will work as a partner on the job. Of course, any time you have an opportunity to partner a job, you'll jump at it. So it's in your best interest to become skilled at giving an effective sales presentation.

In our office, we call them *dog 'n pony shows*. While we don't relish leaving the construction site to try to become public speakers, we know it's necessary. I've found an excellent magazine that can help you with your presentations and provide you with some good ideas on where to find presentation materials. I recommend looking into a subscription. Write or call:

Presentations Magazine
Lakewood Publications
50 S. Ninth St.
Minneapolis, MN 55402
Phone: 612-333-0471
Fax: 612-333-6526

You can also visit their Web site at:

www.presentations.com.

Quality craftsmanship, aggressive completion schedules and cost-efficient construction; these strengths enable Westra Construction, Inc. to deliver the results today's owners expect. We invite you to discuss with us your project's objectives. We will show you how Westra Construction can build it better.

General Contractor

Design/Build Contractor

Construction Manager

CAD/Video Animation

Value Engineering

Service Department/Special Projects

Crane/Equipment Rental

Rigging/Equipment Moving

Precast/Steel Erection

Water/Sewer Underground Utilities

Plumbing

Pre-Engineered Systems

Demolition/New Construction/Renovations

Figure 16-3
Create effective advertising techniques

Standing in front of a group and making a presentation can be quite stressful if you're not used to it. However, there are a few things you do to help calm your nerves and make the event a little easier. Let's review the art of the presentation:

1. First, and most important, know your stuff! Whether you do the presentation or you have someone else do it, make sure that whoever's speaking is totally familiar with your company, the project you're going to be discussing, and the people you'll be talking to.

2. Unlike written ads or brochures, a customer presentation offers you the opportunity to describe your company and your personnel in some depth. Your prospective clients will have a marked interest in knowing about your business and the people with whom they're dealing. They want to know that their investment will be in safe and competent hands.

3. Using an audio/visual aide is always a good idea. It breaks up the monotony of having to look at and listen to one person for the entire presentation. You can use slides, charts, graphs, pictures, photos, renderings or anything else of interest that will help you get your message across. Just be sure that it really does enhance your presentation. Check out *Microsoft PowerPoint®* and other presentation software packages. These packages have a wide array of graphics and features that can make even a novice look polished and professional.

4. When preparing your audio/visual presentation, keep it simple. Don't try to cover eight or nine points with every slide or frame you present. You'll only succeed in confusing your audience. Let the message on the slides present the concept, then fill in the fine points for your audience.

5. The fancy software and hardware can help make a beginning speaker more comfortable in front of a crowd, but don't forget that technology is just a tool. Don't use it as a crutch.

A good speaker can hold his audience's attention and deliver his message with genuine interest and sincerity, even without visual aids. This level of confidence comes with time and experience. If you do it enough, you'll get there.

6. Rehearse, rehearse, rehearse, and then rehearse some more! Contrary to popular belief, *no one can just walk out in front of a group and give a perfect presentation without putting a lot of time and effort into their preparation.* Those polished speakers who make it look so easy weren't just born with that gift — they worked hard to get it down right! Take it from someone who's been there, nothing is more frustrating or embarrassing then messing up in front of a group of people that you're trying to impress. So rehearse!

7. Interact with your audience. Get them involved in your presentation. Ask pointed questions that show you've done your homework. Encourage responses, then listen carefully to their answers. If you can get them talking about the project, it will calm everyone down, and the result will be a better presentation.

8. Get involved in the project. If possible, get a jump on your competition by offering the client some type of unique service. If you were giving a presentation for a new school construction project, for instance, you could offer to help in the referendum campaign. You could help distribute flyers or collect names for the mailing list — anything (legal, of course) that might set your company apart from the pack.

Trade Shows

Concrete workers have *concrete shows*, steel workers have *steel shows*, and glass installers have *glass shows*. These trade shows are everywhere! And they serve a good purpose. For many contractors, trade shows are an efficient and effective means of dispensing and receiving new information on their specialty or the industry in general. It's also a good place to show your wares in front of potential clients.

As a way to advertising your business, however, trade shows come with their own set of risks. For instance, you can lay out a substantial investment for your booth, literature, freebies and staffing, only to find that the show has a far smaller audience than anticipated. Or, the audience that shows up just isn't interested in your display. Here are some things to consider if you think that participating in a trade show may be good for your company:

▮ I prefer to participate in shows where the attendees have to pay to get in. People who have an investment in the time they spend at a trade show are almost always more serious buyers, or at least better leads, than those who just came out to kill an afternoon.

▮ Set clear objectives about what you want to accomplish at the show. Is your goal to solicit leads for your sales staff to follow up on after the show? Or do you want to close sales right there at the show? Are you presenting conventional building ideas or introducing a brand new pre-engineered building line? Focus on what you want to do and then come prepared to follow through.

▮ You can't accomplish your objectives if people don't visit your booth. Offer some enticement. Freebies are one common means of getting people to your booth. You could get an industry expert or a celebrity look-alike to bring in traffic. At one recent show, a participant had a Barney Fife impersonator standing around. It was a little off the wall, I admit, but he did get a lot of people's attention. An industry expert will provide a more serious and informative "documentary" quality to your presentation. People like to be entertained, but they also want to be educated. They appreciate the opportunity to talk with an expert.

▮ See about getting some advance promotion through the show sponsors. Very often, they'll have their own Web site or the show may be featured in a special show issue of an industry trade magazine. Exposure in trade magazines or on the Internet gives prospective customers a chance to check you out before they meet you at the show. Finally, don't forget to take photos of your booth to use in future promotional literature.

Benefits of Trade Shows

The main benefit of participating in a trade show is *exposure*. You get your face and your company name in front of the public. Call it billboarding if you like, but the fact is that we're all influenced by things we see or hear repeatedly. Your appearance at trade shows provides that same effect. If your company name is familiar, it's more likely to come to mind when someone needs the type of service you offer.

Another side benefit of being at a trade show is that you may occasionally have some media exposure. Often, local radio and television station personnel will stroll around the show and conduct interviews with those attending. I've been interviewed about a half-dozen times — and that kind of attention can be an unexpected windfall.

Attending shows gives you a chance to observe your competitors, and check out any new angle they're attempting to play. It also gives you the chance to see if your competition is even *attending* the shows.

Other Ways to Get the Word Out

There are many other ways to get your message out. One important, and rapidly-growing means is via the Internet. We'll discuss that in detail later in Chapter 18. Another is the press release.

Occasionally, you'll have the opportunity to release information about an upcoming project in the form of a *news* or *press release* in a local paper or regional magazine. Try to slip in a little information about your company. These releases tend to be factual in nature, usually just a brief report on the project. But if prepared properly, the news release offers good promotion for your business. Keep it very straightforward and businesslike. No flowery wording or blatant commercialism.

Editors get hundreds of press releases, and they don't like to have their time wasted. If you're sending the news release to their office, it's a good idea to provide a short summary statement at the top. This makes the editor's job a little easier and can result in wider use of the release. You should also find out which editor is in charge of this type of information and submit it directly to that editor.

Getting the Advertising Juices Flowing

What's that? You say you don't have any good ideas for advertising? To get your creative marketing juices flowing, try the following exercise. Ask yourself questions similar to those you asked when you were first beginning to outline your business plan. Remember? They should be something like these:

▌ Are we accomplishing the goals that we originally set for the company?

▌ Have our goals changed due to circumstances that we hadn't anticipated?

▌ Is the company still moving forward, or have we grown stagnant in some areas?

▌ Is the company's growth too slow, or perhaps too fast?

▌ Are we landing the type of work and dealing with the type of customer that we originally set out to reach?

▌ Is it time to modify our goals, set new goals, or take the company to the next level?

By answering these questions, you can get a picture of where you were heading and where you are now. You may find out your goals have changed over time. Then you can focus more clearly on how to market your company to meet those goals. You may have started out planning to do one type of work and found yourself a niche in another area entirely. Sometimes that happens, and you have to adjust your strategy to suit that new market. Once you've established your marketing direction, then it's only a matter of staying enthusiastic and *maintaining your focus*.

Of course, you might not feel comfortable handling marketing and advertising yourself. In that case, you may want to hire a marketing firm to do your advertising for you. They'll charge a hefty fee, but it can be worth the cost to get professional assistance.

Some General Rules for Marketing

Our industry offers up some unique considerations for marketing. Commercial construction offers a product that's extremely expensive. We often forget that. Decisions to buy your services aren't made overnight. Be careful that you don't dismiss prospects too early, even if they don't seem to be taking immediate action on their proposed projects. The decision to go ahead with a project takes time. It's often affected by things that have absolutely nothing to do with your company or product. Perhaps they're arranging for financing, or looking for a better lot on which to build their new facility. Be patient.

Another important rule is to take the time to learn as much as you can about the type of business and the customers you want to pursue. Your clients will be impressed with your knowledge, and your understanding of their needs will allow both of you to move more quickly to the heart of the deal.

Finally, be sure to set aside a percentage of your budget for marketing and advertising. Don't treat it as an afterthought, and don't forget to figure it into your overhead expenses. It should be a continuing part of your business activities. Remember, it takes a long time to get a project moving, so you have to start selling your product way in advance if you're going to keep your crews working. Your marketing program should be an important part of your business. The more seriously you take it, the more seriously your business will be taken.

Chapter 17

The ADA Impact

For many commercial contractors, the *Americans With Disabilities Act*, or *ADA* for short, looms like a dark cloud over their construction projects. ADA regulations are often ambiguous and difficult to apply, and looking into them can leave you with more questions than answers.

Much of the time, however, ADA compliance means just adding a ramp here or a few levered latchsets there. So most contractors comfort themselves with the knowledge that ADA compliance only affects a small portion of their contract. They simply accept this complication in their scope of work (along with the added cost), make the required alterations, and blame it on the government. They never give the ADA more than the briefest necessary consideration.

And who can blame them? It's not like they have time to just sit around and think about all the various rules and regulations that they have to deal with! With the everyday deadlines and pressures of being a commercial contractor, there usually isn't time to argue or debate such matters. Unfortunately though, if we don't question, we don't understand why we must comply or just how much this compliance costs us, both as contractors and taxpayers.

And that's the problem. You see, sometimes complying with ADA means much more than pouring a little concrete or purchasing a few pieces of hardware. Complying with ADA can be very expensive. For example, retrofitting an older two-story downtown building for handicapped access could involve building concrete access ramps, installing handicap doors and hardware (which may require widening door widths), and adding elevators. Construction of an elevator enclosure involves all the items that go along with building an enclosure, plus sump pits (and sewer line), electrical installations, exhaust fans, smoke detectors, telecommunications, etc. With so much detail and so many trades involved, even what may appear to be a straightforward ADA retrofit can easily eat up $100,000 or more, and that's an amount large enough to put a crimp in any project.

The Americans With Disabilities Act

It's important to state right off that *no* reasonably compassionate person is going to argue against the rights of disabled people to have access to the same buildings and facilities that the rest of us use everyday. But there are economic realities that enter into the picture for those of us in commercial construction. Unfortunately, complying with ADA costs money — sometimes, lots of money. And in an industry that's so intensely competitive and driven by economics, the question of whether or not to spend additional funds for ADA faces anyone designing and building a commercial construction project. Compliance may even spell the difference between moving ahead with a project or canceling it all together.

With so much at stake, it indeed becomes important to understand the subject. So, what is the American With Disabilities Act? Who really needs to comply? What's the contractor's obligation in ensuring that the requirements are met? And, what are the penalties for noncompliance? Let's look at the answers to these questions.

Interpreting the ADA

The *Americans With Disabilities Act* sprang to life on July 26, 1990. Sponsored by Senator Thomas Harkin and signed into law by President Bush, the act basically deals with discrimination — and continues a trend that includes the Civil Rights Act of 1964, the Architectural Barriers Act of 1968, the Rehabilitation Act of 1973, and the Fair Housing Amendments of 1988. In general, the ADA addresses architectural barriers, communication barriers, and other cultural encumbrances that may negatively impact disabled people.

Although there are many facets to this act, the one that impacts the construction professional is the one dealing with "architectural barriers." From our point of view, the primary goal of the act is to ensure disabled people the right to employment based on their ability to have *reasonable access to commercial and public facilities.*

The act covers all places of public accommodation, commercial facilities, and government-owned buildings and facilities. It does *not* apply to religious facilities, single- or multi-family housing, Indian tribal properties, or private clubs. The ADA isn't a building code (though portions of the law have been adopted into many state and local building codes), it's a *civil rights act.* Compliance requirements are often determined on a case-by-case basis, interpreted by the individual assigned to hear the case. This means that it's entirely possible that the contractor, architect, engineer, and owner may never really know beforehand whether or not their design will *fully* comply with the ADA. After all, what's the exact legal definition of *"reasonable access?"*

That's where our problems begin. Given this vague definition, it's not at all surprising that nobody wants to take responsibility for compliance, including the designer and the contractor. In bid documents, architectural and engineering firms routinely include clauses that waive any and all responsibility for ADA compliance. (In a 1996 case, the Federal District Court dismissed a case against a Minnesota design firm for ADA violations brought by the Paralyzed Veterans of America. It determined that compliance with the ADA rests with owners and operators, not design professionals.)

If *they're* not going to be responsible, I'm sure not going to be responsible! Why should the contractor be left holding the bag? After wading through accounts of past court cases, all with varying outcomes, it's virtually impossible to come up with any clear sense of who's really responsible for compliance. If you *could* generalize, the responsibility seems to fall the most heavily on the owner — the person who looks to those of us considered building professionals for expert construction advice and council.

Dealing With the Wording

In all fairness, this hesitancy to accept responsibility isn't because we're all a bunch of liability-dodgers maliciously plotting to shirk our duties. The wording of the act is simply ambiguous and difficult to apply. Title I of the ADA *"prohibits private employers, state and local governments, employment agencies and labor unions from discriminating against qualified individuals with disabilities in job application procedures, hiring, firing, advancement, compensation, job training, and other terms, conditions and privileges of employment."*

You're probably just as socially conscious as the next guy, but when you're up to your armpits in alligators, all you really want to know is *whether or not you have to put in a darn ramp!*

The best we can offer you is a reasonable interpretation of the definition of *reasonable* accommodation. As defined by the act, *reasonable (architectural) accommodation* may include, but is not limited to:

▌ Making existing facilities readily accessible to and usable by persons with disabilities.

▌ Acquiring or modifying equipment or devices to be made usable by persons with disabilities.

An employer is required to make accommodations for the known disability of a qualified applicant or employee as long as it doesn't impose an *undue hardship* on the operation of the employer's business. Undue hardship is defined as *an action requiring significant difficulty or expense when considered in light of factors such as an employer's size, financial resources and the nature and structure of its operation.* (Again, that's discernable only on an individual, case-by-case basis.) An employer is not required to lower quality or production standards to make an accommodation, nor is an employer obligated to provide personal use items like glasses or hearing aids.

Common ADA Applications

For commercial contractors, common areas of work that often involve some sort of ADA compliance include:

▌ *Parking facilities:* Curb ramps, access to aisles or main pathways, accessible routes to an equally-accessible building entrance, and accessible parking spaces with signs and markings for handicap vans (normally located as close as possible to the building or facility).

▌ *Ramps:* Wheelchair access ramps to buildings and facilities, with railings and landing areas where ramps change directions. The ramps must meet minimum slope requirements, commonly not exceeding 1:12 with railings, or 1:20 without.

▌ *Stairs:* Special handrail design and extensions as well as minimum depth and width requirements for landings to accommodate (non-wheelchair) handicapped persons.

▌ *Doors:* Meet minimum width requirements for adequate clearance and maneuvering of wheelchairs; provide lever handles in lieu of turn knobs to minimize having to grasp and twist wrist to open doors. Figure 17-1 shows required doorway width and depth clearances.

▌ *Objects protruding into the path of travel:* Permanent objects protruding into the path of travel overhead or from the side must be modified, or the pathway modified, to allow clear circulation and passage for a handicapped individual. See Figures 17-2 and 17-3. Obstructions include posts, columns, fire extinguisher cabinets or electric water coolers (which also have required mounting heights — see Figure 17-4).

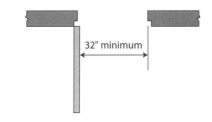

A Hinged door

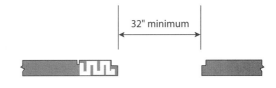

B Folding door

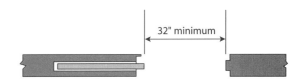

C Sliding door

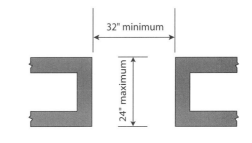

D Maximum doorway depth

Figure 17-1
Clear doorway width and depth

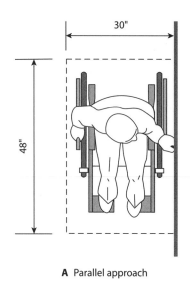

A Parallel approach

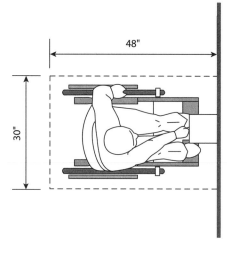

B Forward approach

Figure 17-2
Minimum clear floor space for wheelchairs

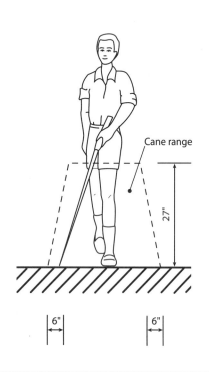

Cane range

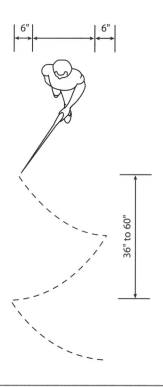

Figure 17-3
Clear space required for cane technique

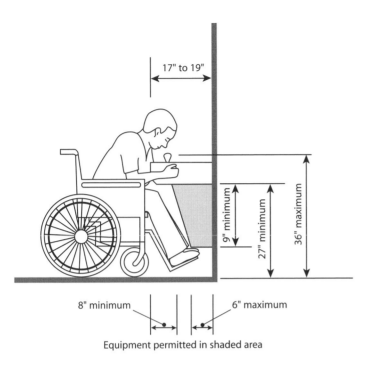

17" to 19"

9" minimum

27" minimum

36" maximum

8" minimum   6" maximum

Equipment permitted in shaded area

A Spout height and knee clearance

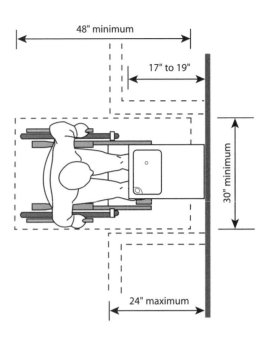

48" minimum

17" to 19"

30" minimum

24" maximum

B Clear floor space

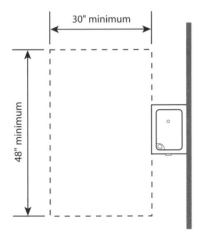

30" minimum

48" minimum

C Free-standing fountain or cooler

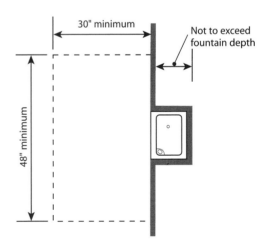

30" minimum

Not to exceed fountain depth

48" minimum

D Built-in fountain or cooler

Figure 17-4
Drinking fountains and water coolers

▮ *Path of travel requirement:* If you're renovating or altering an area of a building, you may be required to make the *path of travel* from outside the building to that area comply with accessibility guidelines. This can include altering parking spaces, entrances, restrooms, telephones, water fountains, and anything else along the path of travel. However, there's a rule, known as the 20 percent disproportionate cost rule, that's designed to ensure that this requirement doesn't force inordinate economic hardship on the owner of the property. If, for example you're renovating an area hundreds of feet from the nearest accessible entrance (as in a large manufacturing plant), the cost of altering the path of travel would probably be disproportionately large. If such alterations cost more than 20 percent of the projected budget for the project, they fall under the disproportionate cost rule.

▮ *Toilet rooms/bathrooms:* Toilets and bathrooms must be accessible. This includes bathrooms located within locker rooms or other employee areas. The doors to enter these spaces must swing *into* the space and meet minimum width requirements. A 5-foot unencumbered turnaround circle is often required in the room, but the door may swing into the space (see Figure 17-5). Toilet stalls, showers, lavatories (see Figure 17-6), and toilet accessories (grab bars, towel bars, dispensers, etc.) all must be at ADA-required mounting heights, and comply with reach requirements, and other criteria. There are also regulations governing signage in the toilet/bathroom areas and the presence of visual alarms for the deaf.

▮ *Signage:* Special signs must be located throughout the building, including all main corridors, bathrooms, and parking areas, indicating handicapped access. These signs must also be in Braille for the visually impaired.

Special Construction

In addition to the general requirements for buildings and facilities, there are other rules that apply to special types of construction. Restaurants must comply with additional regulations involving food service areas and condiment and utensil placement. Hotels must comply with regulations involving width requirements for corridors and access doors into and within guest rooms, roll-in showers, visual alarms in the guest rooms, special wheelchair-accessible rooms, and more. And there are special requirements for children's environments, including day care centers and public play areas.

The Consequences for Noncompliance

As ambiguous as the regulations may seem, there can be costly and serious repercussions for the violator. If a disabled person files and wins an ADA suit, they are generally entitled to a remedy *"that would place them in the position they would have been in if the discrimination had never occurred."* This would likely mean that the building in question would have to be retrofitted with the required architectural accommodations to the satisfaction of the courts. This is often far more expensive than if the accommodations had been included in the new construction in the first place.

The offended person may also be entitled to be hired, promoted, reassigned, reinstated, or receive back pay (or some other remuneration and/or accommodation), depending on the offense. The business owner or employer may also be ordered to compensate the victim for future monetary losses, mental anguish and attorney's fees. If the court decides that the employer acted with malice or reckless indifference, he may be liable to pay punitive damages. Those are real consequences — *and real costs* — that someone has to pay. The suit may be brought against the employer or building owner, but it could also include the builder and the architect.

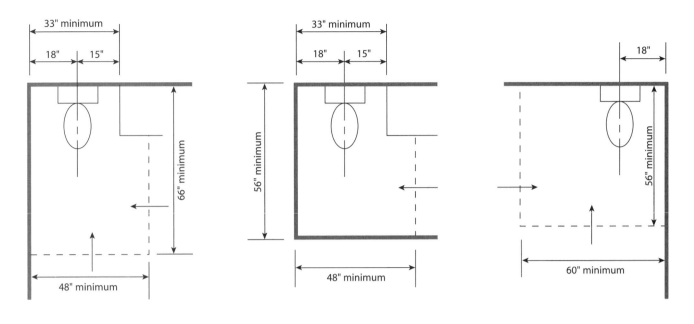

Figure 17-5
Clear floor space for water closets

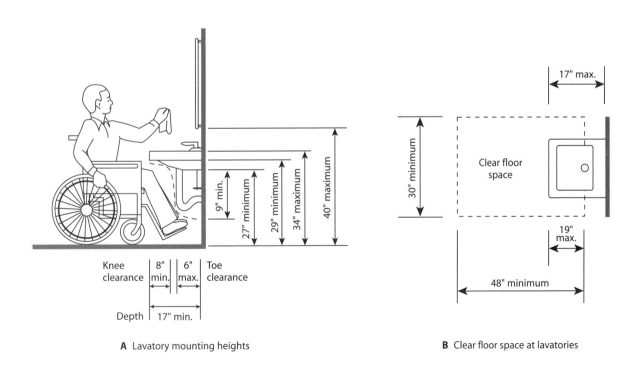

A Lavatory mounting heights

B Clear floor space at lavatories

Figure 17-6
Lavatory clearances and mounting heights

Police Power

Don't be lulled into complacency by the apparent lack of code authority involvement. Though the ADA isn't generally enforced through local building inspectors, it's quite possible that it could be adopted into the code on a state or local level. When in doubt, check with your local code authorities. Also, if the ADA details are included on the drawings, the inspectors can still nip you for plain old, everyday *plan* compliance. Even if the state and local code bodies don't involve themselves, there's plenty of compliance authority out there.

The EEOC

First and foremost is the *U.S. Equal Employment Opportunity Commission (EEOC)*. They're the ones who, on July 26, 1991, issued the regulations that enforced Title I of the ADA (which took effect a year later).

The original doctrine made employers with 25 or more employees comply with ADA. Then, on July 26, 1994, the number of employees went down to 15 or more. Anyone who believes that they're a victim of discrimination, because they were denied reasonable access to or within a particular building or facility, can file a grievance at any field office of the U.S. Equal Employment Opportunity Commission. They have offices located all over the U.S.

The Individual

Secondly, and perhaps most importantly, disabled persons have, with the backing of the court systems and other agencies behind their cause, a police power of their own. They possess the right to file suit at any time. That threat alone is generally adequate to force the contractor, and anyone else involved in the construction of a building, to go ahead and *"put the ramp in."*

My best advice is to educate yourself about ADA regulations, comply with the details on the plans, and ask lots of questions when in doubt. In general, the contractor is fairly low on the liability list. Often, the fact that you've made a sincere effort to comply with ADA regulations will keep the wolves at bay. Use common sense and, of course, never, ever eliminate or ignore an ADA detail on the architectural plans or in the specifications without getting express permission to do so.

Resource Information

If you'd like to know more about the ADA and EEOC, you can request information by calling the EEOC, toll-free, at 1-800-USA-EEOC.

The U.S. Chamber of Commerce publishes a pamphlet entitled *"What Businesses Must Know About the Americans With Disabilities Act."* You can request a copy by calling them at 1-800-638-6582.

There are also excellent resource groups you can contact, such as:

University of Illinois at Chicago
Great Lakes Disability and Business Technical
  Assistance Center (MC 626)
Department of Disability and
  Human Development
1640 West Roosevelt Road
Chicago, Illinois 60608
Phone: 1-(800) 949-4332
            (312) 413-1407
Fax: (312) 413-1856
Web page: *www.gldtac.org*

There's a lot of excellent ADA information on the Internet. The U.S. Justice Department has a site devoted strictly to the ADA. It's located at:

www.usdoj.gov/crt/ada/adahom1.htm

If that doesn't help you, go to any search engine, type Americans With Disabilities Act, and go surfing.

You can get information and architectural guidelines through a variety of other sources as well. These include: the Architectural and Transportation Barriers Compliance Board (1-800-USA-ABLE), building and code organizations, local trade unions, state or local governments offices, vocational and disability training/rehabilitation agencies, the U.S. Attorney General's office, your local library, and more. All it takes is the desire to know, and a little investigation. As you'll see, you're not alone. Help is all around you.

Chapter 18

The Contractor, the Computer and the Internet

Few things since the invention of the hammer have had so profound an impact on our business as the computer. Most of us started out in residential construction with our accounting system on the back of an envelope and our estimating and bidding submittals on the back of our business card — in pencil! But today, computer applications power virtually every aspect of the commercial construction industry, from architecture to accounting. There are new applications emerging almost daily. And computer technology is only going to have a greater role in our industry in the years to come.

This is no time to be timid and stand by while your future slips away from you. If you want to stay competitive in this increasingly-complex field, you absolutely *must* learn to use a variety of computer applications. If you're not computer literate, you'll be at a distinct disadvantage in today's market place. Unfortunately, commercial construction involves many trades, disciplines, and specialized needs, which makes it difficult to choose the computer applications that are right for you and your employees.

Computer Basics

I realize there are probably still a few of you out there who have avoided any computer knowledge so far, and are still able to keep your residential building business going. For you, let's start with a few basics. If you're already computer literate, skip ahead to the section, "The Computer and Commercial Construction."

Hardware

The term *hardware* simply refers to all of the physical and mechanical devices that make up the computer system. Typical hardware items include the computer itself, which is housed in a box and contains the central processing unit (CPU), co-processors, random access memory (RAM), a hard drive, a CD-ROM drive, possibly still a floppy drive, and a modem. Attached to the computer you have a monitor, a keyboard, a mouse, usually a printer, and whatever other peripheral devices you might need, such as plotters, scanners, or digital cameras.

Data is stored on the computer's hard drive. The RAM is the active memory where the data resides. Whenever you open a program and use it, you're using RAM. The more RAM you have, the larger the programs you can operate on your computer. When it comes to RAM, more is definitely better.

Information is processed by the CPU and displayed on your monitor. Complex mathematical calculations may require co-processors. The CD-ROM or floppy drives are used for loading new programs onto your hard drive or for operating programs that aren't permanently stored on the hard drive. The modem allows you to communicate via phone lines with the outside world. You can connect with other computers, send e-mail and access the Internet.

Software

The programs that you load into your computer are called *software*. Although the basic applications will already be factory-loaded when you get your system, you can buy and install additional programs on CD-ROM or floppy disks to suit your individual needs. Word processing, spreadsheets, estimating, scheduling, accounting applications, architectural design, and engineering are all common programs that you may want to use eventually. These programs let you perform specific tasks, which we'll discuss later.

You can master most of these programs with just a modest investment of your time and effort. As your company grows and matures, the software applications you need to accommodate your growth will be a little more complex. You may have to periodically upgrade your hardware to handle larger programs and increased data.

The Computer and Commercial Construction

So what does all of this mean to the commercial contractor? Well, to better understand the role of the computer in commercial construction, we'll break it down into two categories: project design and office applications.

Project Design

There are software application packages that allow architects, builders, designers, engineers, and draftspeople to transfer their thoughts and visions to paper. The most commonly-used design application is a CAD (computer-aided design) program. There are many different types of CAD software available, at costs ranging from a few hundred dollars to many thousands of dollars, especially if you start adding in peripheral items such as plotters, scanners, and digitizer pads to supplement the system. How much you pay for a setup will almost certainly be a product of your need, your pocketbook, and the future benefit you feel you'll get from your investment.

However, this isn't an investment to be taken lightly. You really have to have a need before taking this plunge. If you do a lot of design/build and you're constantly generating sketches and prelimi-

nary plans, the cost will probably justify itself over time. If you seldom draw up your own plans and details, you're better off passing on this investment and paying someone else to do your drawings. I know of builders who've made substantial purchases in CAD programs and peripheral equipment and never used them. They're not easy to learn. It takes considerable time and some technical ability to put these applications to use for you.

A CAD program is a design tool, like a drafting table and T-square, that lets the designer create organized and detailed working drawings (blueprints), details and cross-sections, and specifications for a construction project. Figure 18-1 shows a floor plan for a medical facility drawn using a CAD program. These very large and powerful programs employ computer graphics, drawing tools, denotation and text capability, arrows, numbering systems and other features to create drawings. The original drawings can still require a fair amount of time to produce, maybe as much as comparable ones done by hand, but there's little industry argument that CAD offers considerable time-savings when it comes to altering or making *changes* to drawings.

With CAD, changes are clean and quick. There's no erasing, patching, and redrafting that can leave the final document messy, muddy, and fragmented-looking. Altered drawings come out just as clean and sharp as the originals, and much more presentable to the client. You can also save each set of changes, in case the customer decides to "go back to the way it was," or there's some disagreement about when or why a change was made.

In addition, and perhaps most importantly, CAD offers the designer and the consumer options they never had before. Besides the almost unlimited versatility, the designer now has the capability to demonstrate a design concept or idea in *three-dimensional* format. They don't have to build models from balsa wood and paper any more! This 3-D capability, which gives viewers the feeling that they're walking through a room or down a corridor, offers a clearer picture of the final architectural product. You can't achieve this kind of perspective with two-dimensional drafting. You can bet that it didn't take long for designers and builders to dis-

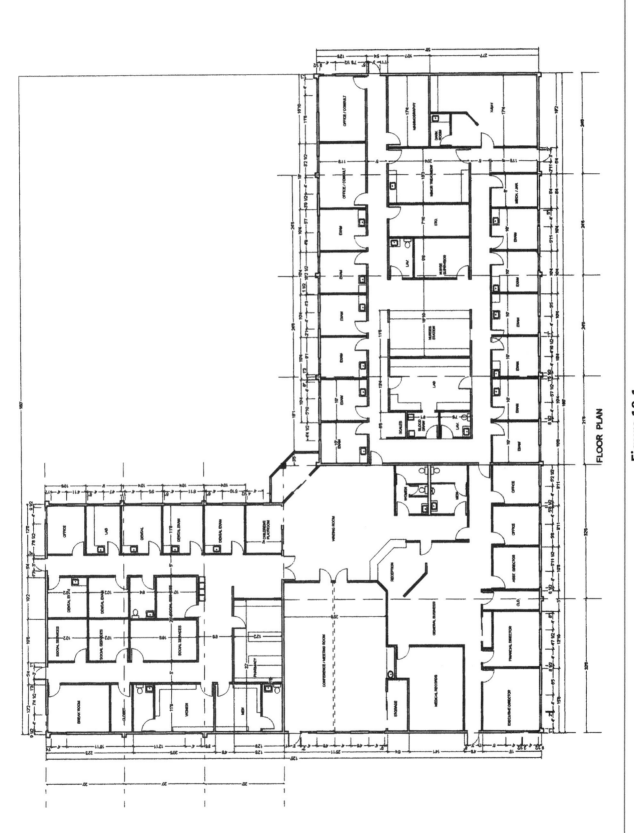

Figure 18-1
Floor plan

226 Moving to Commercial Construction

cover the benefit of using this 3-D feature as an effective sales tool when presenting their ideas to clients.

Aside from this powerful marketing feature, when CAD programs are integrated with engineering and design packages, they also allow designers and engineers to create, lay out, and test common construction design elements. Roof or floor trusses, beams, columns, posts, modular wall or building assemblies, horizontal and vertical precast concrete panels, plumbing, mechanical and electrical designs and schematics, and more, can be tested under computer-simulated load and stress conditions. Computer testing these elements saves time, money, effort, and may even save an employee or consumer from potential harm by uncovering unforeseen problems in a structural component.

There are also software packages that allow the engineer/designer to perform tasks such as:

▍ Determine projected energy calculations/use for a new building or facility.

▍ Audit and analyze energy use in *existing* buildings.

▍ Determine air tightness, air movement and indoor air quality in new or existing buildings.

▍ Design computer-generated prototypes or "mock-ups," eliminating the expense of building real ones.

▍ Test and balance mechanical and electrical systems.

▍ Create computer models of daylight and/or artificial light schemes for buildings.

▍ Allow different interior finishes and color schemes to be reviewed by decorators and clients.

The Future of Computer Design

What's next in construction computer design technology? My guess is that it'll come in the area of *intelligent* or *smart* buildings. You've probably been hearing these terms for a while now. The development of this technology — though slow and awkward at first — is beginning to come to fruition. In

fact, we've seen it in many of the items we've just discussed, but it will get even more comprehensive. It employs various elements of computer science to make it work, including communications technology and *artificial intelligence (AI)*. *AI* allows the computer to answer questions and even make "judgments" within its particular area of expertise.

With intelligent design, a building or facility will be able to automatically:

▍ Optimize energy use throughout the building, including sending heat or air conditioning only to those areas of the building where the computer "knows" workers or visitors are present.

▍ Offer heating and cooling alternatives to small zones within a building, such as individualizing the heating/cooling needs within an office or a worker's cubicle.

▍ Monitor, regulate, and correct indoor air quality and air flow.

▍ Allow operation of a building's facilities management program from remote locations.

▍ Adjust lighting levels throughout a building.

▍ Monitor and police visitors and employees throughout the facility.

Office Applications

Once a project is through the design stages, it has to be built! Let's look at some software applications that provide practical help for the commercial contractor.

Word Processing

Although word processing programs aren't designed specifically for construction, I listed this software application first because it's surely the one you'll be using the most. I know I do. Letters and communications are a critical part of the commercial construction process, so a good word processing program is worth its weight in gold. *Microsoft Word*® and *Corel WordPerfect*® are a couple that I've used, but there are many other good packages on the market.

A word processing program works like a typewriter, but has many more functions and is much more convenient. Besides writing business letters, you can use your word processor to create databases of clients, subcontractors and suppliers. Then you can have these names automatically inserted in a flyer or form letter, and print them out. The program will also address the envelopes, which you run through your printer as well. In addition to generating letters and correspondence, these programs allow you to electronically store your client communications. That can save you hours of hunting for letters you wrote to clients months ago.

Figures 18-2 and 18-3 are samples of how you can use a word processing program to set up a form letter and a mail merge list to conveniently correspond with your subs and suppliers. The program will automatically insert the names and addresses of those listed on the merge file into the merge letter and address an envelope. I create a database for each job just for this purpose. The software programs come with instructions and tutorials to help you set your files up for various types of correspondence.

The newer word processing programs allow you to perform many functions that in the past were only offered with desktop publishing programs. You can create pamphlets, flyers, promotional materials, banners, even your own letterhead — and you can print them out in color. They also come with a library of writing aids, such as a spelling and grammar checker. I really appreciate this. Of course a weakness in spelling doesn't mean I can't put up a quality commercial building, but some customers may think that sloppy, semi-literate correspondence is a sign of less than expert work. I'm a great fan of the templates for creating business documents, and there are other features that can make the novice businessperson look much more professional.

Spreadsheets

Microsoft Excel®, *Corel Quattro Pro*®, and *Lotus 1-2-3*® are a few popular spreadsheet programs. I've used all of them from time to time, and they're excellent. They have a wide range of features and presentational formats and come with detailed instructions. For the commercial construction professional, there's no better way to handle large quantities of numbers than with a computer spreadsheet. I've often wondered how construction estimators used to manage all those calculations by hand. They're constantly receiving and altering subcontractor and supplier proposals, especially on frantic bid days, and having to calculate and recalculate the numbers! The error rate must have been phenomenal. Or maybe those guys were just a whole lot smarter.

> *"For the commercial construction professional, there's no better way to handle large quantities of numbers than with a computer spreadsheet."*

Figure 18-4 shows a sample spreadsheet used to compare bids from six subcontractors. The bids shown are incomplete. As the final numbers come in, you just plug them in and the program will total them for you. You don't have to worry about errors in addition or last-minute hurried calculations. As long as you plug in the right numbers, you'll get a correct total.

Besides letting you calculate quickly and accurately, you can use these programs to create tables, graphs and charts that you can use in client presentations. Good spreadsheet programs are available at most computer stores at reasonable prices.

Estimating and Project Management

Some of the names I've come across in commercial estimating or project management programs include *National Estimator*®, *MEANS*®, *MC²*®, and *Timberline*®. There are others as well, designed for different trades. These programs can save you considerable time over manual estimating methods, and help you organize your estimates. Estimating programs may be a little harder to find than common spreadsheet packages because they're more specialized. While some are quite reasonable, others can run into a substantial investment, especially when you add in digitizer tables and other peripherals. Whether you go for the simpler types or the more expensive ones, you'll find they're worth looking into.

Phone	(555) 888-8888	JAMES GILBANK
Fax	(555) 888-8889	THOMAS GILBANK
		GARY GILBANK

GILBANK CONSTRUCTION, INC.
GENERAL CONTRACTORS
Commercial   Industrial   Residential
200 MAIN STREET   GORE, WISCONSIN 55555

Established 1965

April 13, 2000

FIELD(name)
FIELD(address)
FIELD(city & state)
FIELD(phone number) , FIELD(fax number)

Attn: FIELD(attention)
Re: Value-engineering ideas for the Lienwaiver Inn

Dear FIELD(salute)

There will be a pre-construction meeting for the Meyer Inn construction on April 20th, 2000 @ 2 PM at the jobsite located at 500 Peasant Road, Janesville, WI. Please have a company representative there as we will be discussing schedule, safety, and other pertinent start-up topics. Thanks!

Sincerely,

GILBANK CONSTRUCTION, INC.

Steve Saucerman

cc: file

Figure 18-2
Sample merge letter

FIELDNAMES(name;address;city & State;phone Number;fax Number;attention; salute;csiDivision)ENDRECORD

Bass & Sons ExcavatingENDFIELD
7000 Canyon RoadENDFIELD
Janesville, WI 55555ENDFIELD
608-755-2345ENDFIELD
608-755-2346ENDFIELD
Tom BassENDFIELD
TomENDFIELD
2100ENDFIELD
ENDRECORD

Brothers Asphalt PavingENDFIELD
2900 Longhorn DriveENDFIELD
Janesville, WI 55555ENDFIELD
608-755-9999ENDFIELD
608-755-9998ENDFIELD
Joe BrothersENDFIELD
JoeENDFIELD
2150ENDFIELD
ENDRECORD

Prairie Dog LandscapeENDFIELD
300 Hwy 16ENDFIELD
Vernon, WI 53333ENDFIELD
262-777-2223ENDFIELD
262-777-2224ENDFIELD
Steve AdamsENDFIELD
SteveENDFIELD
2900ENDFIELD
ENDRECORD

Figure 18-3
Sample merge file

Rough wall comparisons - Meyer Inn (Material only — FOB jobsite)

	Williams	Stark	Lenear	Mindant	Hosta	Automo
Basement (stick)	7 ,256 *panelized	5,643	Incl blw	No bid	No bid	Pending
Main level (stick)	22,169 *panelized	38,588 *panelized	162,723	No bid	No bid	Pending
Subtotal (Bsmt & Main stick)	29,425	44,231	162,723	-	-	-
2nd floor (panels)	30,513	38,588	Incl abv	No bid	162,628	Pending
3rd floor (panels)	26,909	38,588	Incl abv	No bid	Incl abv	Pending
4th floor (panels)	26,909	38,588	Incl abv	No bid	Incl abv	Pending
Additions	4,414					
Subtotal (2nd, 3rd, 4th)	88,745	115,764	-	-	162,628	-
Floor trusses	29,311	34,042	51,634	26,554	Incl abv	28,800
Roof trusses	13,868	12,416	28,062	16,678	Incl abv	17,976
Subtotal (trusses)	43,179	46,458	79,696	43,232	-	46,776
3rd & 4th floor sheathing	14,224	15,928	Incl abv	No bid	Incl abv	Pending
Roof sheathing	5,924	8,462	Incl abv	No bid	Incl abv	Pending
Subtotal (sheathings)	20,148	24,390	-	-	-	-
Subtotal - misc. framg/blkg	11,864	12,201	4,541	-	-	Pending
Grand totals	**193,361**	**243,044**	**246,960**	**43,232**	**162,628**	
				N/A	N/A	N/A
Notes:	gyp applied		gyp applied			

Figure 18-4
Sample spreadsheet

The CD at the back of this book has the *National Estimator* program, plus 300 pages of labor and material costs for commercial construction. This is the full program, not a demo. To try it out, just insert the CD in your computer. This program, along with much larger databases for residential, electrical, painting, remodeling, plumbing & HVAC, is available from Craftsman for less than $50, including a book. I've used many different programs, and found this one to be well conceived and comfortable to use.

Scheduling

A few popular scheduling programs are *PrimaVera SureTrack®*, *Timeline®*, and *Microsoft Project®*. These programs allow you to create and present construction progress schedules for use by workers, managers, and clients. You can track target completion dates and milestones, and compare them to actual dates. They allow you to monitor and track your jobs in a variety of ways, including breaking down the schedules according to trade, task, phase, or other criteria. That lets you view the schedule from varying perspectives. You can even assign manpower requirements to specific schedule items. The programs also offer a variety of formats for presentation; you can add legends, tables, headings, and detail boxes, and print your final schedule in color, which makes it easy to read and understand.

Job Accounting, Job Costing, A/P and A/R

There are many excellent accounting software packages on the market, some specifically designed for construction accounting, such as *Timberline®*, *Concord CS2000®*, *Peachtree®*, *QuickBooks Pro®*, and more. The CD in the back includes a handy little program called *Job Cost Wizard*, that turns your *National Estimator®* estimates into invoices that you can export directly into QuickBooks. That makes it easy for you to invoice your jobs, and do job costing.

Next to word processing, these are the programs that probably get the most use in the commercial construction office. They guide the office accountant/bookkeeper through the accounting process, and often include the ability to link different accounting modules, like job cost to A/P, to save time and provide a check on accuracy.

Miscellaneous Software

As your company grows and your needs change, you may decide to expand your software and hardware to include other applications. You can get software that monitors various operations, such as tool and equipment monitoring (with bar-code readers), or inventory tracking. There's also presentation software, such as *Microsoft PowerPoint®*, for formal sales presentations; communications software for faxing; video conferencing software, such as *Microsoft NetMeeting®*; Internet online access and e-mail programs, among others.

Don't worry about having to buy and learn all of these programs right from the start. The process is a gradual one, and you'll know when it's time to make the next move. After you've had time to assimilate all this information, take a moment and catch your breath before we look at what's almost certainly the next great leap in construction communications and applications: *the Internet!*

The Internet

I'll let you in on a little secret: Once you have access to the Internet, it won't be long before you'll get the itch to go online and surf. The Internet has revolutionized communications — and construction marketing. It levels the playing field for the contractor. The Internet allows smaller or newer firms, who generally can't afford to put a lot of money into marketing and advertising, to compete economically in a larger market. Companies can collect customer information, perform market research, conduct surveys, and get their message out to more potential customers, at a fraction of the cost of more conventional marketing and research methods.

Then there's e-mail. E-mail is something that you need to use on a regular basis to really appreciate. I have to admit that when I first began using e-mail I was kind of skeptical about it. *"What's the big deal?"* I wondered. Now, five years later, I send and receive a lot more e-mail than letters. It's faster, more convenient, and *it's fun*. You don't have the frustration of playing incessant phone tag. They get your message when they return to their desk, and you get their response when you return to yours. And you'll soon find yourself putting more things "in writing." Then you print a copy of your e-mail to keep for your records — you know how important it is for a contractor to keep good records. It also saves you the cost of long distance phone calls, as you can access the Internet on a local phone number. Then spend as long as you like sending messages to almost anywhere in the world, without any charge at all. And that can save you big bucks in long-distance calls over the course of a year. Because of its convenience, it won't be long before all your clients will expect to communicate with you via e-mail.

Your Web Page

I'm sure you already know something about Web pages. Every day on TV, you hear advertisers, news networks, and even prime time sitcoms trumpet their Web addresses, which look something like this: *www.youshouldbereadingabook.com*. The Web address simply tells you how to access their Web site, also called their home page, on the Internet.

A Web page can be used to share information or promotional material with others. You can use it to tell customers all about you, your company, your services, and how you can help them. You can set up your Web page so potential customers can read the information and then *interact* with your company via e-mail, filling out a form online, or by conventional mail or telephone.

Setting Up a Web Page

Does this sound too scary and high-tech? Well, don't be intimidated. Once you start visiting other company's Web pages, you'll see the advantage of having a page or several pages of your own. They aren't difficult to set up. Believe me, if I can do it, *anybody can!* There are several different Web page software programs that you can use, including *Microsoft FrontPage*® or *Corel WebMaster*® *Suite*. These programs will guide you, step-by-step,

through the process, and explain how you can use different fonts, graphics, backgrounds, and even photos to make your page more appealing to your audience. You'll have a surprisingly professional-looking page (or pages) in a very short time.

The programs also explain how to link your page to other Web pages, and how to incorporate interactive features, like direct e-mail. Once you've created a page that you're happy with, then you have to upload it onto the Net.

> *"Setting up a web page. Does this sound too scary and high tech? Well, don't be intimidated."*

Before you can live in cyberspace, you first have to register your page and online name (called your *domain* name). InterNIC is the principal domain name registration authority. You can locate them at *www.InterNIC.net*. To save you a little of the confusion that I went through, I should tell you that there's also a company called Internic Software (located at *www.internic.com*), a domain name brokerage company that charges a fee to help with domain name registration. Depending on your needs, you may find it money well spent. There are lots of other sites you can go to for help as well. Any search engine will take you to more than you can handle. Here are a few more:

> *http://networksolutions.com*
> *http://Website-hosts.com/domain.html*
> *http://www.us-domain-name-registration.com*

When you select your online name, don't be too cute or too abstract! Choose a name as close to your company name as possible. I say as close as possible, because your first couple of choices may already be taken. You want to make it easy for your customers to find you. If they have to guess, they'll get tired of looking and go elsewhere — and that's hardly the way you want your advertising to work! And make sure to include your Web address on all your printed material, including letterhead and business card.

You will also either need to provide your own *file server* (which is a computer large enough to power your site — and that can be costly), or contract an *Internet Service Provider (ISP)* who'll run your Web site off of one of their servers for a monthly fee. They may charge you an initial setup fee as well. There are many ISPs out there. You might need to do a little investigating to find one that provides the service you want at the price you want to spend. Check *Internet Magazine* or search online.

Once you're registered and on the Internet, you'll want to submit your Web site to several *search engines*. These are the search programs that help Internet users find their way around the Net. Your ISP may provide this submission service for you, but if you're registering on your own, you can submit your site to a number of engines through services such as *Submit-it* (*www.submit-it.com*), for one low fee. *Submit-it* was advertising that they could register you with up to 400 different search engines — and that'll help anybody find you!

Getting Help — If this all sounds like it's over your head, there are a couple of ways to get help. One is to contact your community college and find out if they have any online classes on Web page design and setup. Many offer six-week courses that are inexpensive and easy to follow. The best thing about these classes is that they're taught via the Internet so you can do them at home, at your own pace. You'll get a lot of support and some very good tips on where to find free information and setup help on the Net.

You can also pay to have an independent Web page service company create and publish your page for you. Of course, the more information you want to put on the Web, the more your initial cost and the higher your monthly service fees will be to maintain and update your site. However, for many, it's just easier to pay someone else to provide these services for you.

Web Page Tips

Once you're on the Net, it's simply a matter of keeping your page current and responding to all the requests for information that your page generates. You may find that you'll have to try a few different approaches before you get the kind of the results you're hoping for. To help you add a little spice to your page, here are a few tips for keeping your home page interesting:

▎ Update your page on a regular basis; keep your logo and format consistent, but make changes in content, pictures or background — anything to keep it fresh looking. Nothing's worse for a Web visitor than to see the same old thing every time they come to view.

▎ Your home page should have an index of any additional pages you have on your site. That way, the viewer will have an idea of what's available and where he can go to find it. Keep in mind that most people just have an average-size monitor, not the newer large screens, so don't make your page too wide. Also, they'll probably have to scroll down to read all the information on your page, so if there's a message you *want to get across right away*, make sure it's on the first frame of your page. An ideal location is the upper left-hand corner as the page comes up.

▎ If you have links to other pages on your site, or from your page to other Web sites, be sure to check them periodically to make sure they're working properly — or to be sure that the other link addresses haven't changed. It's frustrating to click on a site that won't come up. It can make your page seem unmanaged and unprofessional, and that's not the image you want to project.

▎ Don't fill the page with a lot of self-serving fluff or boring company biographies. You need to get your customer's attention. Use some fun or interesting graphics or pictures and have a special offer, a promotion, a seminar, an open house, or a tip-of-the-week — some type of new information — that they can read about. Don't go overboard on the graphics, though. Graphics take a long time to download, and if you have too many, or they're too large, it'll take so long to download the page that no one will wait. They'll just move on to your competitor's more interesting, faster-loading site. How about offering a *chat room*, where a group of users can get together online to discuss construction topics in an open forum? Or perhaps you can have an interview with an industry expert, or even have that expert be a part of the chat room discussion? Use your imagination.

▌ Search out and find online news or topic groups that discuss your trade or type of service (there are many construction and contractor groups). If you think it's appropriate, try to get your page listed and linked with them. Also consider links to other related sites, such as building material suppliers, or vendors. Most sites include an e-mail address at the bottom for correspondence and you can request permission to link to these sites — and to have them link to yours. This can give your site added credibility, and often makes the site more interesting to the viewer.

▌ Purchase ad space, often called banners, on other sites. These banners will link that page to yours. One click and the customer's on your home page!

▌ Make sure the body of your text contains key phrases like *design/build, partnering, commercial general contracting,* etc. Many search engines will locate your page based on matching these terms to searched phrases, even though they don't appear in your title or heading.

▌ Incorporate a *counter* in your page. This is a software mechanism that tells you how many people are visiting your page. It's a valuable bit of information that can help you to determine the effectiveness of your page and the search engines bringing in your viewers.

▌ Always remember to include an address and phone number on your page to encourage further contact. Also put in your fax number, e-mail address and the names of people to contact. This gives the company a more *grounded* feel. I always like to have a physical address as well as a Web address for anyone I'm dealing with. I'm sure I'm not the only one who's uncomfortable when I don't have any way to contact the company other than through the Web site. It simply feels suspicious to me — like maybe they don't really exist.

Internet Information for Your Business

The Internet and online communications will most certainly continue to grow in the future. Besides a marketing arena for your company, the Internet offers an inexhaustible supply of up-to-the-minute information on contracting and the construction industry. There are thousands of sites dealing with building, construction and related industries. You can locate most of them using common search engines, such as *Hotbot*®, *AltaVista*®, *Lycos*®, *Yahoo*®, or *InfoSeek*®. Just to get you started, I've compiled a short list of good construction-related sites that I've found helpful. They usually have links to other sites, so in no time you'll find more data than you knew existed! Here's my list:

1. *www.abuildnet.com* — A great all-around site for building and construction resources and general information on the construction industry.

2. *www.aecinfo.com* — Architecture, engineering, construction and home building discussion forums, products, firms, services, news, events, projects, articles, classifieds, specifications, schools and more.

3. *www.askmac.com* — Put out by the Aberdeen Group, this site sells itself as the largest Web site for the concrete and masonry construction industries.

4. *www.build.com* — A directory for building and home improvement products and information. Be sure to explore HomeTalk, Build.com's building and home improvement bulletin board/discussion forum.

5. *www.buildingteam.com* — Building and construction-related information on codes, products, projects, news, liability, management and more.

6. *www.buildingweb.com* — Bills itself as the Internet Center for the Construction Industry.

7. *www.constructionnet.net* — A comprehensive directory of builders, contractors, material suppliers, and design professionals.

8. *www.contractors-license.org* — Information on obtaining a contractor's license in every state in the U.S.

9. *www.craftsman-book.com* — How-to manuals for professional builders, contractors and tradesmen and cost guides for construction estimators and property loss adjusters. You can browse through their books, place orders, and even download trial versions of construction cost estimating databases. Plus there are many links to other useful sites.

10. *www.get-a-quote.net* — Costs from the largest free construction estimating database on the Web, the same data used in the *National Estimator*® estimating reference manuals.

11. *www.onbuilding.com* — An interactive construction planning and management resource dedicated to helping building owners, architects, engineers and contractors improve construction planning and implementation.

12. *http://search.construction.com* — An online directory of construction businesses and building product manufacturers.

What's next? Well, your guess is a good as mine, but I'd say computer technology, the Internet and online business will continue to rocket skyward. If you don't want to be left behind, you'd better climb on board!

Chapter 19

Success and Failure in Commercial Construction

Well, we're almost done. But, before you dive headlong into commercial construction, there's something you need to know. Although commercial contractors may have "graduated" from residential construction to go "play with the big boys," we still screw up, just like everybody else. The difference is that we're usually in higher-ticket projects, so a screw-up hurts more. In fact, just one can finish you!

Every day I watch commercial construction contractors unwittingly undermine their profit margins by neglecting to do the things that will make their businesses stronger and more successful. Even more painful is that I was one of them, and sometimes I forget for a minute and join their ranks once more. In the beginning of this book, I told you that we'd just be discussing the *correct* way to operate a commercial construction business. But now I'm going to reverse that. We're going to take a few minutes to look at how many contractors get it all wrong!

My hope is that by identifying the problems and pitfalls that have tripped me (and many others) during our commercial construction careers, you'll have an easier road to follow. Why walk into the same crocodile-infested swamp after I've already warned you about it? Of course, contractors are an independent lot, and some of us will still go wading in. But keep a copy of this chapter around as a reminder. Every once in a while, review the ten items that I'm about to cover. Then, if you notice yourself heading into one of these hazard zones, maybe, just maybe, you'll be able to catch yourself in time to avoid real damage.

Ten Dumb Things We Do to Spoil Our Own Success

So what are these hazards that we have to be so wary of? Here's my top ten list:

1. We agree to schedules that are unrealistic.

2. We agree to profit margins that are too low.

3. We try to be the owner's friend.

4. We try to be the architect's friend.

5. We accept people at their word.

6. We fail to evaluate the impact of change orders on our bottom line.

7. We try to solve all of the owner's problems.

8. We allow multiple punch lists.

9. We fail to protect ourselves from partisan contracts and documents.

10. We go into commercial construction without proper preparation.

Now let's go over them one by one.

One: We Agree to Schedules That Are Unrealistic

Here's a situation that we've all had to deal with. "Very late in a long and hectic commercial construction negotiation process (after all of the budgets, drawings and specs, hard costs, fees, and contract jargon have been hashed out dozens of times), the final completion date is slipped into the conversation. It isn't what you've assumed it would be. They've moved it a full three weeks ahead! You voice a concern, but the owner comes back with, *"The negotiation process went a little long . . . but I still have to be in by this date!"* So what do you do? You're sure not going to trash the job at this stage — you've got far too much time and money invested! But you can see that the owner is very serious about the date. So you reluctantly accept this change in plan, and *hope you can work it out down the road.* You know it's unrealistic (and deep down you know you can get burned), but you go ahead anyway."

> *"You may lose a few jobs along the way, but in the long run you'll be ahead as a result of keeping your profit levels consistent."*

Don't fall into that trap! *Stand your ground.* You have an agenda and the owner has an agenda. His is no more important than yours. Negotiation is a two-way street, and though it often seems that the owner is holding all the cards (the project itself), he needs you as much as you need him. Don't cave into the pressure of breaking ground. Stop the talks and address the issue right then and there. Take it from someone who's been in that very spot; there's never going to be a better time to have it out than at that moment. If you genuinely believe the schedule is unrealistic, don't wait until 10 days before the completion date to say so. There may be some tense moments and it may even back up the start date further, but it has to be done. For many people, this type of negotiation is a skill that takes time to develop, but you've got to start someplace. You owe it to your business, yourself, your employees, and to the project owner — who may not even know that what he's proposing is unrealistic!

Two: We Agree to Profit Margins That Are Too Low

Recently, while closing out a project, I absent-mindedly left a folder containing all my in-house estimate breakdowns, including my profit and overhead numbers, material proposals, and subcontractor quotations, at the office of the architect on the project. Was I ever embarrassed! I felt I'd let my deep-est and darkest secrets out of the bag. I waited for the call from the architect expressing his indignation that I was taking such advantage of the owner by figuring such an outrageous profit into the job. Well, I did receive a call the next morning. But instead of a reprimand, the architect, hardly able to contain his laughter, snickered into the phone, *"Is this really all you guys make on these jobs?"* I think I would have felt better if he *had* accused me of being greedy!

This episode just reinforced an impression that I'd had for quite some time. Our industry has grown *disproportionately* competitive, and we're now working too hard for too little. Rapid competitive growth, owner sophistication, and architect/owner-biased contracts and project control documents have eroded the ability of the average commercial contractor to enjoy healthy and consistently-sustainable profit margins. Even good jobs can end up unprofitable due to elements that are simply out of our control.

It's been my observation that the trap of diminishing profit creeps up on us. A strongly competitive market has a way of slowly and stealthily eroding our competitive urges. Here's a tip: periodically stop and examine your profit margin. This is so important that you must mark a date on your calendar — *and do it.* Then evaluate your findings. Have you fallen into a pattern of letting the market dictate your percentages? Are you being forced into margins that are too low to sustain your overhead and foster growth? If so, an adjustment is in order.

You may lose a few jobs along the way, but in the long run you'll be ahead as a result of keeping your profit levels consistent. You may also find that there's some advantage in not taking every job that's dropped on your doorstep. A less hectic workload may allow your office and field staff to operate more efficiently, increasing your overall productivity. And increased productivity translates into enhanced profit margins. It takes discipline to stick to this exercise instead of letting it fall by the wayside. You must recognize that it takes more than skill and hard work to succeed. It also takes good business practice and planning.

Three: We Try to Be the Owner's Friend

There's nothing wrong with developing a cordial relationship with the owner over the course of a construction project. It's a normal and natural evolution resulting from your prolonged association. However, should you actually become friends with the owner, you may find that your relationship is often at odds with the toughness you need for sound business reasoning. In other words, you're more likely to give away items of work that you should charge for simply because he's your friend.

Don't forget that your first relationship with the owner was as a businessman, and you must retain that businessman's state of mind. Since you're providing a valuable service, it's only fair that you charge for that service. Believe me, the owner would certainly have to pay someone else for it. Never feel guilty about charging fairly for your work.

Four: We Try to Be the Architect's Friend

We're also likely to develop a friendly relationship with an agreeable architect — and that can be potentially even more perilous than becoming friends with the owner. The architect is generally far more involved in the day-to-day workings of the project. When you're dealing with a friendly architect, it can be very easy to fall into a pattern of *give-and-take*, swapping a current favor for a future favor. The trouble comes at the end of the job, when all of the favors suddenly come down to money. The architect, once again the owner's staunch representative, develops a severe and debilitating case of *"selective memory."* The promises (that weren't necessary to put into writing) that would have provided you with compensation for all the additional work you did? Forgotten. Even more incredible is the fact that he *can* manage to recall, with crystal clarity, *every credit* that should be issued back to the owner.

Don't let friendship cloud your judgment. When it come to contracts and change orders — ***get everything in writing***. Don't count on promises — there's no profit in it.

Five: We Accept People at Their Word

I've noticed that, as a group, we contractors seem to have a tendency to take a man at his word. Many of us still embrace a lot of the builder's old-school, golden-rule type philosophy, perhaps handed down from our fathers and grandfathers. This includes the idea of sealing a deal with a handshake and accepting verbal commitments. When you say you're going to do something, you do it. End of story. And you assume that others do the same. We don't want to demand that everything be in writing — especially those *"gentlemen's agreement"* kinds of things.

But, as we discussed earlier, we live in a different world from that of our fathers and grandfathers. Most of us have felt the sting of being too trusting more than once. *"Next time will be different!"* you vow. But, on the very next job, you fall right back into the same old pattern and *you get burned again*.

At some point you have to begin to ask yourself if your reason for accepting verbal promises is really noble, or *just an excuse for being lazy in the field*. Is it just a justification to cover up your failure to record costs and administer changes the way they should be done? Ask yourself that question. The answer may be hard to swallow, because you *know* that you need to get everything in writing — no excuses!

Six: We Fail to Evaluate the Impact of Change Orders on Our Bottom Line

Mishandled change orders, particularly changes in the field, have an *enormous negative impact* on our profit line item. Changes in the scope of work occur on every construction project, and the effective handling and administration of these changes is *critical* to maintaining a profitable job. The problem is, change orders can nick you in so many different ways. It's truly a difficult task to address them in every aspect of the job.

What do I mean by "every aspect of the job"? Well, let's look at just a few of the ways we let the change order process suck profit from our pockets.

Ignoring Small Changes

It's so easy to let the small changes go. No one wants to be thought of as penny-pinching, or petty. These are the items that seemed *"not so big at the time"* or *"not worth the paperwork to write up"* — things like running the base molding a little further than detailed, or digging the trench a little deeper than was called out on the plans. Of course, it's no secret that a whole lot of small things add up to a whole lot of dollars at the end of the project. And all that money is unaccountable — and your loss!

Giving In to Promote Public Relations

There's always the temptation to "throw in" a change for an owner who has become a friend, or an owner that you want to impress because he may offer you future work. Though I could make a case for an occasional gesture if you think there's a genuine opportunity for additional work, the important thing is *not to get carried away* with public relations generosity. Commercial contracting is a business, and most owners understand the concept of charging for your product or service. After all, they're in business, too.

Making Changes Without Getting an Adjustment in the Schedule

There are some changes that don't affect the construction schedule, but there are many more that call for an extension of the project completion date. Asking for the additional cost to cover the change is often as much as many contractors want to do, especially if they think they can "fit it into the schedule." But what happens when they're wrong? What about the liquidated damages that they'll have to pay at the end of the job? The owner and architect won't be shy about demanding this penalty, so don't you be shy about demanding an extension to the schedule if it's called for by a change.

Starting a Change Before It's Costed and Agreed On

On most commercial construction projects, there's a constant, pounding pressure to maintain the schedule. This is in spite of the fact that the owner and architect are making changes that, in the spirit of

cooperation, you're trying to make in as congenial a manner as you can muster. However, keep in mind that once you perform that extra work, *your negotiating position has been greatly weakened*. When the job is completed, you may be forced or coerced by the owner and architect (who are holding onto your money) to accept whatever additional amount they feel is appropriate for that extra work — an amount that can range from reasonable to ridiculous. Once again, if you *always* get it in writing *before* you do the work, you can be pretty sure you'll get paid! Another advantage of strictly adhering to this policy is that the number of changes seems to diminish. Could it be that when the owner and architect see that changes are going to cost *them* time and money, instead of just you, the changes become less necessary?

Failing to Charge for Office and Administrative Time

We often fail to charge adequately for the additional office and administrative time required for processing change orders. I recently completed a city hall project that required 250 billed manhours to cost, negotiate, account for, and administer the change orders — *and that was just for change orders!* At $50 an hour, which is a conservative office billing rate, the administrative cost of the changes alone came to $12,500. There were hundreds of changes on this job, and the architect argued about the cost for each one. At the end of the job, we estimated that we'd put twice as many manhours into administering changes than all the other office work required for the job, including bidding. That was *real* time and *real* money, and it would have been a real loss to the company if we hadn't charged for it!

Failing to Cost and Bill Change Orders Carefully

Changes can come fast and furious, and it's all too easy to just take that additional quote for louvers from the HVAC sub, mark it up, and shoot it out as a pricing proposal for change. But wait a minute. Did you include general requirements in your price? General requirement increases are a legitimate part of any change (sometimes even on change credits) and should always be taken into account.

You should track the material, labor, and equipment required for each change separately on the job site. Make sure time cards and material tickets break out the changed work, and that the information makes it to accounting. And finally, make sure that the changes are billed and collected. You have to follow through on these items. I've seen change orders that were processed just fine right up until the time they were to be added to the continuation sheet for billing, *and then they never made it!*

Seven: We Try to Solve All of the Owner's Problems

Many of us decided to move up from tradesmen to contractors because we felt we had a knack for solving problems. We're mechanically- and technically-oriented people who enjoy focusing on and solving a variety of tasks. This is a wonderful attribute that serves us well in our profession, but it can also get us into trouble. You see, as problem solvers, we often have a tendency to express our opinions on just about any matter where we see a need — and it seems we're *always* seeing a need. On the job site, this tendency shows itself in our desire to correct what we see as flaws, errors or missteps in the construction and design of a project. The trouble begins when the owner listens to what we say (and we often say way too much!) — and then the change orders begin to roll!

The good news is that everyone's entitled to his opinion and we're allowed to speak our minds. The bad news is that *it's not our job to design the building!* That's what architects and engineers are for. So my advice is to keep your mouth shut and let them do their jobs. Even if you think they aren't doing their job, that's a problem that should be worked out between the owner and architect. Don't try to solve all the owner's problems, because that's what they are: the *owner's* problems.

Every time we open up a "can of worms," the project schedule grinds to an unceremonious halt, and that affects our labor productivity and our profit. When labor slows, money is lost! So, keep quiet and keep the schedule moving. Just be content in the knowledge that you're secretly *soooooo* much smarter than everybody else on the job site.

Eight: We Allow Multiple Punch Lists

I'm not sure where or when the idea of multiple punch lists came about, but I'm sure a contractor wasn't there when it happened. This practice pretty much throws fair play right out the window, allowing the architect to generate one punch list after another, each one with new items that didn't even exist when you completed the first list. I've had jobs with three, four and even five punch lists. And, because the architect is holding your money, you can't disregard them. At one point on a job where the number of lists became ridiculous, I simply announced to the owner and architect: *"We are done!"* Then I pulled my men, and demanded my money. My mistake, of course, was that I should have done this after we completed the *first* list.

Make sure everyone understands from the start that there will be just one punch list — period!

Nine: We Fail to Protect Ourselves from Partisan Contracts and Documents

Let me take this moment to point out something that you may not know: *commercial contractors rarely read the specification manuals cover to cover.* Surprised? No? I really didn't think you would be. These manuals have grown too large and too complex; there just isn't time to read them through completely. In fact, I'd be a little concerned about a contractor who *did* have the time to read the manual all the way through! However, this is a genuine problem. With the exception of a small minority of obsessive-compulsive construction-contract enthusiasts, these manuals rarely get read, comprehended, reviewed, and, if need be, rebutted by contractors.

Because it's relatively easy for owners, architects, and engineers to incorporate weighted and partisan wording into the specifications they present to us, knowing that no one is going to go through it carefully, they've shifted more and more responsibility and liability onto the contractor. When you read through one of these manuals today, you'll find that contractors are left with comparatively few legal rights and/or leverage on the very projects that they're building.

Take the time to really read Division 1 of a large specification manual sometime and you'll see what I mean. If taken literally, there's virtually nothing from which the owner and architect *aren't* absolved. Classic clauses go something like this:

"Although the designer has made every effort to ascertain that information contained in the working drawings and specifications is correct, it is the sole responsibility of the contractor to verify and comply with all field and site conditions, code authorities, governing bodies . . . "

Now throw a few *"for no additional compensation"* and *"the architect's interpretation shall be final"* clauses into the pot, add a shovel full of boiler-plate specifications, and you've whipped up a foul-tasting, but very effective owner/architect-indemnity stew!

So, when disputes do come up, our occasional legal victories are more a product of defiant stands, moral indignation, and favorable rulings on ambiguous wording than on actual detailed rights. Having few tools to work with, we're often left to extreme retaliatory measures such as work stoppages, liens, or legal actions. Legal actions are generally only a threat — remember, we've entered into *their* contract, so victories are few and far between. But most businesses lose time and money going to court, so the threat has some weight. It's in everyone's interest to not let it go that far.

Our best defense is to be on the offensive and never allow disputes to get out of hand. Read your contracts, try to read as much of the specification manual as you can, contest items that are blatantly unfair, and make sure all changes are documented.

Ten: We Go into Commercial Construction Without Proper Preparation

Regardless of the difficulties, residential contractors still want to move their operations up into the ranks of commercial construction. Unlike you, many don't prepare themselves properly for this transition. Just by reading this book, you're a step ahead of most of these contractors. And that's good! But don't stop here — do as much research as you can on commercial contracting and business practices in general, and *then* take the plunge.

Good luck!

What's on the CD

The CD-ROM inside the back cover has blank copies of the forms in this book, saved in six formats. Install these blank forms to your hard drive in the format or formats you prefer. Then change the forms to meet the needs of your business. You can fill out the forms on screen or print the forms in blank and fill them out by hand.

The table on the next page lists these forms. Note the file naming convention. The file name for all six versions of the Bid Log begin BIDLOG. The file name extension (the last two or three letters after the period) is different for each version. The extension identifies the program which will open and display the form on your computer screen. Microsoft *Word* for *Windows*™ and Microsoft *Word* for Mac forms have an extension of DOC. Microsoft *Excel* for *Windows*™ and *Excel* for Mac have an extension of XLS. *WordPerfect* files have the extension WP. *Works* files have the extension WKS. To use any of these forms, you'll need one of these six programs.

By default, all 42 forms are installed to your computer in four formats: *Word* for *Windows*™, *Excel* for *Windows*™, *WordPerfect* and Works. To change this default, select "Custom" during the installation process.

Also on the CD is Craftsman's estimating program, *National Estimator*, including *Show Me*, a video demonstration, and *Job Cost Wizard*, a program that exports estimates to *QuickBooks Pro*.

To start using the CD, insert the disk in your CD drive and wait a few seconds. A short video should start automatically. When the video is done, you'll be invited to click "Install Software." When installation is done, the *National Estimator* program will begin. (If the video doesn't start automatically, click **Start**. Click **Settings**. Click **Control Panel**. Double-click **Add/Remove Programs**. Click the **Install** button and follow instructions on the screen.)

To use any of the blank forms, click **Start**. Click **Programs**. Click **Construction Estimating**. Click **Moving to Commercial Construction Forms**. Click on the form of your choice. The form will open if you have the program required to view that form. By default, forms are installed to the C:\My Documents\M2CC Forms folder and can be opened by browsing to that location.

If you use a Mac, browse to the Forms directory on the CD and click the Excel_Mac or Word_Mac folder. Windows users can also open forms in the Forms directory on the CD. But you'll have to save forms to your hard drive if you make any changes.

Form name	Book page number	File name
Bid Log	75	BIDLOG
Change Order	144	CHNGORDR
Change Order Log	145	CHNGLOG
CSI Estimator	62	CSIEST
Customer Satisfaction Survey	174	SURVEY
Daily Construction Log	133	DAYCONLG
Daily Equipment Log	135	DAYEQLOG
Daily Material Log	138	DAYMATLG
Deficiency Notice	111	DEFNOTE
Delay Notification	140	DLAYNOTE
Employee Data Sheet	28	EMPDATA
Employee Exit Interview	158	EXIT
Employee Turnover Tracking Report	151	EMPLOYEE
Estimate Summary Sheet	79	ESTSUM
Fax Bid Request	77	FAXBID
Fax Cover Sheet	26	FAXCOVER
Field Authorization for Change	147	FLDCHNG
General Requirements Worksheet (Administrative)	48	ADMINREQ
General Requirements Worksheet (Field)	46	FLDREQ
Interoffice Memo	24	INTEROFF
Job Invoice	178	JOBINVCE
Lien Waiver	170	LIENWV
Memo	25	MEMO
Percentage Complete Invoice	179	PERCTINV
Plan Log	19	PLANLOG
Project Closeout Checklist	172	CLOSLIST
Project Closeout Letter	168	CLOSLETR
Project Punch List	164	PUNCH
Purchase Order	181	PO
Quotation (bid)	81	BIDQUOT
Safety Agreement	199	SAFEAGR
Safety Checklist	198	JOBSAFE
Safety Meeting Report	200	SAFEMEET
Statement (balance forward)	177	BALFWD
Subcontractor Agreement	125	SUBCONTA
Subcontractor Comparison Sheet	117	SUBCOMPR
Submittal/Shop Drawing Log	110	SUBMTLOG
Supplier Comparison Sheet	118	SUPCOMPR
Time and Material Invoice	182	TMINVCE
Transmittal Form	27	TRANSFRM
Unit Estimate Worksheet	78	UNITEST
Weekly Equipment Summary	136	WKEQLOG

Index

Practical References for Builders

Construction Forms & Contracts

125 forms you can copy and use — or load into your computer (from the FREE disk enclosed). Then you can customize the forms to fit your company, fill them out, and print. Loads into *Word* for *Windows*™, *Lotus 1-2-3*, *WordPerfect*, *Works*, or *Excel* programs. You'll find forms covering accounting, estimating, fieldwork, contracts, and general office. Each form comes with complete instructions on when to use it and how to fill it out. These forms were designed, tested and used by contractors, and will help keep your business organized, profitable and out of legal, accounting and collection troubles. Includes a CD-ROM for *Windows*™ and Mac. **400 pages, 8¹/₂ x 11, $41.75**

How to Succeed With Your Own Construction Business

Everything you need to start your own construction business: setting up the paperwork, finding the work, advertising, using contracts, dealing with lenders, estimating, scheduling, finding and keeping good employees, keeping the books, and coping with success. If you're considering starting your own construction business, all the knowledge, tips, and blank forms you need are here. **336 pages, 8¹/₂ x 11, $28.50**

Contractor's Guide to the Building Code Revised

This new edition was written in collaboration with the International Conference of Building Officials, writers of the code. It explains in plain English exactly what the latest edition of the *Uniform Building Code* requires. Based on the 1997 code, it explains the changes and what they mean for the builder. Also covers the *Uniform Mechanical Code* and the *Uniform Plumbing Code*. Shows how to design and construct residential and light commercial buildings that'll pass inspection the first time. Suggests how to work with an inspector to minimize construction costs, what common building shortcuts are likely to be cited, and where exceptions may be granted. **320 pages, 8¹/₂ x 11, $39.00**

Land Development

The industry's bible. Nine chapters cover everything you need to know about land development from initial market studies to site selection and analysis. New and innovative design ideas for streets, houses, and neighborhoods are included. Whether you're developing a whole neighborhood or just one site, you shouldn't be without this essential reference. **360 pages, 5¹/₂ x 8¹/₂, $62.00**

National Construction Estimator

Current building costs for residential, commercial, and industrial construction. Estimated prices for every common building material. Provides man-hours, recommended crew, and gives the labor cost for installation. Includes a CD-ROM with an electronic version of the book with *National Estimator*, a stand-alone *Windows*™ estimating program, plus an interactive multimedia video that shows how to use the disk to compile construction cost estimates. **616 pages, 8¹/₂ x 11, $47.50. Revised annually**

Blueprint Reading for the Building Trades

How to read and understand construction documents, blueprints, and schedules. Includes layouts of structural, mechanical, HVAC and electrical drawings. Shows how to interpret sectional views, follow diagrams and schematics, and covers common problems with construction specifications. **192 pages, 5¹/₂ x 8¹/₂, $14.75**

Contractor's Survival Manual

How to survive hard times and succeed during the up cycles. Shows what to do when the bills can't be paid, finding money and buying time, transferring debt, and all the alternatives to bankruptcy. Explains how to build profits, avoid problems in zoning and permits, taxes, time-keeping, and payroll. Unconventional advice on how to invest in inflation, get high appraisals, trade and postpone income, and stay hip-deep in profitable work. **160 pages, 8¹/₂ x 11, $22.25**

CD Estimator

If your computer has *Windows*™ and a CD-ROM drive, *CD Estimator* puts at your fingertips 85,000 construction costs for new construction, remodeling, renovation & insurance repair, electrical, plumbing, HVAC and painting. You'll also have the *National Estimator* program — a stand-alone estimating program for *Windows*™ that *Remodeling* magazine called a "computer wiz." Quarterly cost updates are available at no charge on the Internet. To help you create professional-looking estimates, the disk includes over 40 construction estimating and bidding forms in a format that's perfect for nearly any word processing or spreadsheet program for *Windows*™. And to top it off, a 70-minute interactive video teaches you how to use this CD-ROM to estimate construction costs. **CD Estimator is $68.50**

Commercial Metal Stud Framing

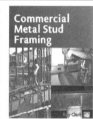

Framing commercial jobs can be more lucrative than residential work. But most commercial jobs require some form of metal stud framing. This book teaches step-by-step, with hundreds of job site photos, high-speed metal stud framing in commercial construction. It describes the special tools you'll need and how to use them effectively, and the material and equipment you'll be working with. You'll find the shortcuts, tips and tricks-of-the-trade that take most steel frames years on the job to discover. Shows how to set up a crew to maintain a rhythm that will speed progress faster than any wood framing job. If you've framed with wood, this book will teach you how to be one of the few top-notch metal stud framers. **208 pages, 8¹/₂ x 11, $45.00**

Residential Steel Framing Guide

Steel is stronger and lighter than wood — straight walls are guaranteed — steel framing will not wrap, shrink, split, swell, bow, or rot. Here you'll find full page schematics and details that show how steel is connected in just about all residential framing work. You won't find lengthy explanations here on how to run your business, or even how to do the work. What you will find are over 150 easy-to-read full-page details on how to construct steel-framed floors, roofs, interior and exterior walls, bridging, blocking, and reinforcing for all residential construction. Also includes recommended fasteners and their applications, and fastening schedules for attaching every type of steel framing member to steel as well as wood. **170 pages, 8¹/₂ x 11, $38.80**

Basic Engineering for Builders

If you've ever been stumped by an engineering problem on the job, yet wanted to avoid the expense of hiring a qualified engineer, you should have this book. Here you'll find engineering principles explained in non-technical language and practical methods for applying them on the job. With the help of this book you'll be able to understand engineering functions in the plans and how to meet the requirements, how to get permits issued without the help of an engineer, and anticipate requirements for concrete, steel, wood and masonry. See why you sometimes have to hire an engineer and what you can undertake yourself: surveying, concrete, lumber loads and stresses, steel, masonry, plumbing, and HVAC systems. This book is designed to help the builder save money by understanding engineering principles that you can incorporate into the jobs you bid. **400 pages, 8¹/₂ x 11, $36.50**

Commercial Electrical Wiring

Make the transition from residential to commercial electrical work. Here are wiring methods, spec reading tips, load calculations and everything you need for making the transition to commercial work: commercial construction documents, load calculations, electric services, transformers, overcurrent protection, wiring methods, raceway, boxes and fittings, wiring devices, conductors, electric motors, relays and motor controllers, special occupancies, and safety requirements. This book is written to help any electrician break into the lucrative field of commercial electrical work. Updated to the 1999 *NEC*. **320 pages, 8¹/₂ x 11, $36.50**

Getting Financing & Developing Land

Developing land is a major leap for most builders - yet that's where the big money is made. This book gives you the practical knowledge you need to make that leap. Learn how to prepare a market study, select a building site, obtain financing, guide your plans through approval, then control your building costs so you can ensure yourself a good profit. Includes a CD-ROM with forms, checklists, and a sample business plan you can customize and use to help you sell your idea to lenders and investors. **232 pages, 8¹/₂ x 11, $39.00**

Construction Estimating Reference Data

Provides the 300 most useful manhour tables for practically every item of construction. Labor requirements are listed for sitework, concrete work, masonry, steel, carpentry, thermal and moisture protection, doors and windows, finishes, mechanical and electrical. Each section details the work being estimated and gives appropriate crew size and equipment needed. Includes a CD-ROM with an electronic version of the book with *National Estimator*, a stand-alone *Windows*™ estimating program, plus an interactive multimedia video that shows how to use the disk to compile construction cost estimates. **432 pages, 11 x 8¹/₂, $39.50**

Estimating Excavation

How to calculate the amount of dirt you'll have to move and the cost of owning and operating the machines you'll do it with. Detailed, step-by-step instructions on how to assign bid prices to each part of the job, including labor and equipment costs. Also, the best ways to set up an organized and logical estimating system, take off from contour maps, estimate quantities in irregular areas, and figure your overhead. **448 pages, 8¹/₂ x 11, $39.50**

CD Estimator Heavy

CD Estimator Heavy has a complete 780-page heavy construction cost estimating volume for each of the 50 states. Select the cost database for the state where the work will be done. Includes thousands of cost estimates you won't find anywhere else, and in-depth coverage of demolition, hazardous materials remediation, tunneling, site utilities, precast concrete, structural framing, heavy timber construction, membrane waterproofing, industrial windows and doors, specialty finishes, built-in commercial and industrial equipment, and HVAC and electrical systems for commercial and industrial buildings. **CD Estimator Heavy is $69.00**

Contractor's Guide to QuickBooks Pro 2001

This user-friendly manual walks you through QuickBooks Pro's detailed setup procedure and explains step-by-step how to create a first-rate accounting system. You'll learn in days, rather than weeks, how to use QuickBooks Pro to get your contracting business organized, with simple, fast accounting procedures. On the CD included with the book you'll find a QuickBooks Pro file preconfigured for a construction company (you drag it over onto your computer and plug in your own company's data). You'll also get a complete estimating program, including a database, and a job costing program that lets you export your estimates to QuickBooks Pro. It even includes many useful construction forms to use in your business. **304 pages, 8¹/₂ x 11, $45.25**

Building Contractor's Exam Preparation Guide

Passing today's contractor's exams can be a major task. This book shows you how to study, how questions are likely to be worded, and the kinds of choices usually given for answers. Includes sample questions from actual state, county, and city examinations, plus a sample exam to practice on. This book isn't a substitute for the study material that your testing board recommends, but it will help prepare you for the types of questions — and their correct answers — that are likely to appear on the actual exam. Knowing how to answer these questions, as well as what to expect from the exam, can greatly increase your chances of passing. **320 pages, 8¹/₂ x 11, $35.00**

Contracting in All 50 States

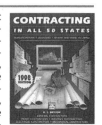

Every state has its own licensing requirements that you must meet to do business there. These are usually written exams, financial requirements, and letters of reference. This book shows how to get a building, mechanical or specialty contractor's license, qualify for DOT work, and register as an out-of-state corporation, for every state in the U.S. It lists addresses, phone numbers, application fees, requirements, where an exam is required, what's covered on the exam and how much weight each area of construction is given on the exam. You'll find just about everything you need to know in order to apply for your out-of-state license. **416 pages, 8¹/₂ x 11, $36.00**